Extracting the Future

Extracting the Future

LITHIUM IN AN ERA OF
ENERGY TRANSITION

Mark Goodale

UNIVERSITY OF CALIFORNIA PRESS

University of California Press
Oakland, California

Cataloging-in-Publication data is on file at the Library of Congress.

ISBN 978-0-520-40278-2 (cloth : alk. paper)
ISBN 978-0-520-40279-9 (pbk. : alk. paper)
ISBN 978-0-520-40280-5 (ebook)

Manufactured in the United States of America

GPSR Authorized Representative: Easy Access System Europe, Mustamäe tee
50, 10621 Tallinn, Estonia, gpsr.requests@easproject.com

34   33   32   31   30   29   28   27   26   25
10   9   8   7   6   5   4   3   2   1

As always, for Romana, Dara, and Isaiah
And to the memory of JCMB—veritas lux mea

CONTENTS

ILLUSTRATIONS

Legend:
- —··— National border
- —·— Department border
- Body of water
- Salar (salt flat)

AMAZON

Madeira

Urubamba

Abuná

Pando

Mamoé

BRAZIL

PERU

Itenes

Cuzco

Beni

ANDES

BOLIVIA

AMAZON

Lake Titicaca

La Paz

La Paz

Arequipa

ALTIPLANO

Cochabamba

Cochabamba

Santa Cruz

Santa Cruz de la Sierra

Tacna

Oruro

Oruro

Sucre

Salar de Uyuni

Potosí

Pacific Ocean

Potosí

Chuquisaca

Tarija

Tarija

PARAGUAY

CHILE

ALTIPLANO

Salar de Atacama

Antofagasta

Salta

Salar del Hombre Muerto

ARGENTINA

ANDES

N

0   100   200 miles
0   100   200 km

Gulf of Mexico

Atlantic Ocean

Caribbean Sea

PERU

BRAZIL

BOLIVIA

Pacific Ocean

CHILE

ARGENTINA

MAP 1. Bolivia. Prepared by Ben Pease.

MAP 2. Potosi Department. Prepared by Ben Pease.

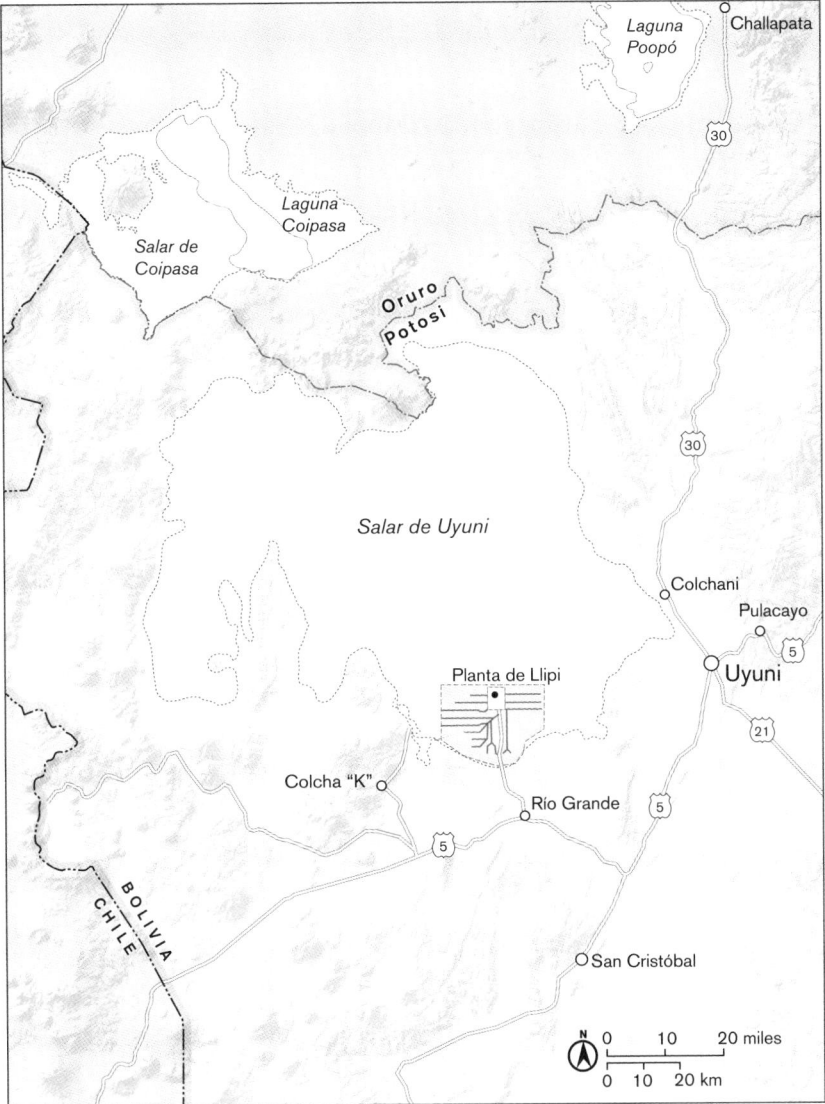

MAP 3. Salar de Uyuni. Prepared by Ben Pease.

Introduction

LOCATING LITHIUM

ON A LATE MORNING in August 2022, I climbed to a sweeping overlook with Alberto Colque, an important community leader from the town of San Cristóbal, which is located in the far southwest of Bolivia's Potosí Department about seventy miles from the border with Chile. I had wanted to get the best view of the full scope of San Cristóbal Mine, one of the largest open-pit silver, zinc, and lead mines in the world. Like all of Bolivia's biggest and most profitable mines, San Cristóbal Mine is owned by a private transnational company, in this case the Sumitomo Corporation, one of Japan's venerable *sogo shosha*, highly diversified global trading companies that emerged after the Second World War.

If resource extraction can be said to have a particular aesthetics, the sight—and visceral experience—of a mature and correspondingly immense open-pit mine deserves its own category. From the observation platform at fourteen thousand feet above sea level, I gazed out across massive mining earthworks, in which thousands of benches, or levels, had been systematically cut into the fragile altiplano terrain, creating multiform and multicolored striations that covered an area the size of a small city. Far below, I could just make out the movement of what appeared to be dozens of tiny vehicles; as I later learned, these were actually cartoonishly giant mining trucks, Caterpillar 797s with thirteen-foot tires, 4,000-horsepower diesel engines, and the capacity to haul four hundred tons of rock up and down the great gashes in the earth.

The observation point had been installed by the mining company in the spirit of the peculiar global ethos according to which largescale extraction projects are assumed by those who build and finance them to be a source of awe and envy. The idea, according to Alberto, was that the mine would come to be seen as one of the wonders of the extractivist world: a place that would draw visitors from near and far, tourists who would travel all the way to this remote and historically forgotten corner of Bolivia to experience a global marvel of engineering, mining geology, and tightly controlled labor. As it turned out, however, the overlook received very few visitors; the well-built, enclosed wooden tower had long fallen into disuse—its door chained shut, most of its windows shattered, traces of adolescent mischief strewn across the dusty floor.

Although I had come with Alberto to look at the mine, he really wanted to show me something else. At the same place, one of the highest and most prominent in the region, is another structure, one that is much older and, for the local Indigenous and peasant communities, of vastly greater importance. Alberto led me into a rectangular enclosure that didn't face the mine but rather the altiplano, which stretches to the horizon to the north and northeast and beyond. All along the waist-high stone walls were large flat rocks meant to be used for sitting. In the middle of the enclosure was a raised table, also built of stones. As Alberto explained, this place was of the utmost significance. Each year, during the season of Todos Santos (celebrated on November 1), community members from the surrounding villages and towns make the climb to the enclosure, where a special *cabildo*, or council, takes place.

During this cabildo, as in many other parts of highland Bolivia, ritual specialists are called upon to make a series of offerings, and as is often the case, the purpose is quite specific. We were visiting the enclosure in mid-August, the last months of the long altiplano dry season, a period of crystal-clear blue skies, cold nights, and, most critically, no rain—the bringer of life. From August through the end of October, the land becomes increasingly parched; the herds of local camelids, like domesticated llamas and wild vicuñas, which are protected by national law, get more emaciated and more desperate to find the already meager grazing plants and shrubs, including the nearly nutritionless *Paja brava*; and community members, almost all of whom incorporate some mix of agropastoralism into their family economies, begin to worry about whether or not the dry season will eventually give way, especially given the commonplace rumors about changing weather, the signs of which seem to be everywhere.

The fact that the mountaintop cabildo takes place during the season of Todos Santos is not a coincidence. Throughout highland Bolivia, Todos Santos is one of the most important periods in the ritual calendar. In the days leading

FIGURE 1. Communal stone enclosure, Nor Lípez province, 2022. Photo by author.

up to November 1, families prepare food and drinks, especially *chicha*, or corn beer, to welcome the spirits of their ancestors, who join the living for a day of feasting, music, and exuberant celebration, most of which takes place inside the village cemetery. The reason the dead are feted with such generosity is that Todos Santos constitutes one important moment in a critical process of reciprocal exchange: once well fed and honored by their living relatives, the dead then ensure that the rains finally arrive and the planting season can begin.

But while the communities of Nor Lípez—the province that includes San Cristóbal—likewise receive the spirits of the dead during Todos Santos in various local cemeteries, they also participate in the special cabildo, which adds an extra layer of reciprocal exchange at the precise moment in which the arrival of rain marks the starkest of boundaries between death and life. Standing with his back to the mine while gesturing to the great expanse of altiplano below, Alberto said that *lipeños* come to this place to make offerings because of its height, which is much closer to the sky than the cemeteries below. Most importantly, llama fetuses are burnt at an elevation at which their ashes are most likely be carried far and wide; as he put it, it is essential that every part of the llama fetus is turned to ash, without leaving any flesh behind.

What the villagers ask for during the cabildo is more than simply rain for their crops, their animals, and themselves. Or rather, the rain is understood to be a consequence of something else. First and foremost, they gather at the enclosure to entreat the winds. Just as on that day in mid-August, the prevailing winds during the dry season come from the west, bringing dust, desiccation, and the smell of the great Atacama Desert, the driest place in the world, parts of which have never received any measurable precipitation since formal meteorological recordkeeping began in the colonial period.

So while the ashes from the burnt llama offering swirl around them, the villagers implore the west wind to stop, to turn back, to retreat to the arid desert from which it comes; and, at the same time, they plead with the winds from the north and northeast to come and bring towering clouds saturated with moisture from the Amazon. If they perform the offering during the cabildo properly, all the ashes from the burnt llama having been taken up into the atmosphere, and assuming all the other ritual offerings of Todos Santos have reached the spirits of the dead as usual, the villagers will be rewarded with rain and the numerous intertwined cycles of life—and death—will continue to rotate in their course.

But while we both stood without speaking as we looked toward the direction from which life itself would, with luck, soon return, something else was clearly visible, something else that—like the strangely beautiful and not-so-strangely horrifying open-pit mine behind us—was also enmeshed in a complex web of climate, materiality, resistance, and longing. What stretched in front of us was another zone of extraction, one that is meant to be part of a radically different future—even though it is in fact deeply connected to the gaping silver, zinc, and lead of San Cristóbal and, more generally, Bolivia's centuries-long history of extractivist entanglement. Moreover, this is a zone of extraction that is understood by locals like Alberto Colque to be firmly embedded in their own local histories of production and community self-reliance, even while its exploitation—like that of the mine—is tightly controlled by powerful and remote institutions.

Looking due north from the enclosure one can make out the transition zone where the dull brown of the altiplano starts to give way to a staggering white. This the beginning of the southern edge of the Salar de Uyuni, the largest *salar*, or evaporated lake, in the world. And although they can't be easily seen from that distance, at least not without binoculars, the southern Salar de Uyuni is also home to a number of industrial facilities controlled by Yacimientos de Litio Bolivianos (YLB), Bolivia's state lithium company. As it turns out, less than thirty miles away from the vantage point of the stone enclosure at which villagers reaffirm their dependence on forces over which they can only ever exercise a fragile agency, another struggle with the elements is taking place, a struggle with global implications.

The highly saline fluid, or brine, that flows beneath the Salar de Uyuni's halite crust contains high concentrations of different minerals, including potassium and magnesium. But it is one mineral in particular, lithium, that has made this desolate corner of Bolivia the object of global economic desire; the proving ground for both technological novelty and repeated frustration; and the site of numerous national, regional, and local social mobilizations. Lithium has also transformed the salar into an elusive landscape that inspires utopian visions of an energy transition in which fuel-burning cars and trucks are replaced everywhere with electric vehicles, or EVs, powered by lithium-ion batteries, made with Bolivian lithium. For among the many other superlatives that attach to the Salar de Uyuni, one in particular crystallizes the stakes—and it is the reason that Bolivia finds itself yet again at the center of a global resource geopolitics whose contours and power lie far beyond its political economic control.

This is the fact that the brine of the Salar de Uyuni, and the brine of the southern salar in particular, contains the largest known deposit of lithium in the world. At the same time, unlike the lithium-rich brines that flow beneath the salares of Chile and Argentina, the other two countries that—with Bolivia—constitute the so-called Lithium Triangle, Bolivia's brine remains largely untouched. Despite modest efforts at small-scale production of lithium carbonate, the compound used in lithium-ion batteries, almost all of Bolivia's lithium remains where it has always been—flowing through a subsurface geology of infinite, fractal complexity.

But without access to Bolivia's lithium, and at megaindustrial scales that far outstrip current well-established industrial production in places like Australia and Chile, it will be much more difficult to supply enough lithium to manufacture the batteries needed for a global EV revolution. And without the capacity to replace fuel-burning cars and trucks with EVs, all of the major energy transition plans, from the €1 trillion European Green Deal (2020) to the different national green energy policies, become simply very expensive blueprints for an impossible future. In other words, so much, to paraphrase William Carlos Williams, depends upon Bolivia's lithium, hidden beneath a surface so white, beside the fields of quinoa.

Yet what does it really mean to say that so much depends upon Bolivia's lithium, which must be extracted, processed, and put into global circulation as a key part of a wider assemblage of energy policymaking, research and development, resource Realpolitik, and ideological confrontation—not the least within Bolivia itself, where the fraught lithium project is unfolding against various histories of extractivism and resource conflict? This book seeks to answer this question by examining the ways in which Bolivia's lithium project is entangled with broader economic, political, and technological processes that define the ongoing and deeply contradictory energy transition, while at the same time viewing these wider contexts through their cultural, material, and historical specificities.

It is impossible to understand the nuances of Bolivia's troubled efforts to industrialize the extraction and processing of its all-important lithium reserves without also understanding the ways in which these efforts are shaped by global energy and climate mitigation policymaking, competition among the major capitalist powers over critical minerals, and histories of resource colonialism and neocolonialism. Yet equally, it is impossible to understand these global forces without tracing the ways in which they are also shaped by both complicated institutional and political dynamics within Bolivia and the lived experiences of the wide range of people and communities who are caught up—in many different ways—with Bolivia's state-controlled lithium project. Indeed, this fact reinforces an argument repeatedly made by anthropologists of energy and environmental justice: that the wider energy transition, which demands the adoption of supposedly green technologies like wind power, solar power, and EVs, is framed as a coherent response to planetary level transformations, including the climate crisis, while what might be called the *actually existing energy transition* is made up of economic, social, and political realities that are necessarily fragmented and—taken together—deeply heterogeneous.

Extracting the Future is an exploration of this fundamental tension based on ethnographic and interdisciplinary research that sought to keep these divergent scales in play over the course of four years that coincided with a number of wider crises and developments, including a short-lived but traumatic coup d'état in Bolivia; the global Covid-19 pandemic, which struck

Bolivians particularly hard; and dramatic changes in the global lithium market, both economic and technological. These crises served to further intensify the interest in Bolivia's reserves, an interest that has coalesced into a potentially ruinous obsession, which epitomizes so much of what is contradictory and self-defeating about dominant global responses to the climate crisis. At the same time, as will be seen throughout the book, this study of Bolivia's lithium project against a background of energy transition privileges the narratives, histories, and perspectives of a wide array of people and institutions who are enmeshed with lithium and its contested promises of transformation.

Finally, much like the view from the stone enclosure, this book offers a number of vantage points onto Bolivia's lithium project and its relation to the global energy transition, vantage points that are paradoxically both far-reaching and necessarily limited. As the anthropologist James Clifford once insisted, ethnography produces a kind of truth about social phenomena—and, perhaps, about the ethnographer herself—that is always partial, always itself entangled in multiple scales of knowledge and ideology, always brushed with shades of ambiguity, but which is nevertheless potentially powerful, even urgent. In this sense, *Extracting the Future* does not offer either a complete history of Bolivia's lithium project—which will, in any case, still be unfolding for years to come—or a readily adaptable guide to how the energy transition might be better structured.

Instead, it presents an ethnographically nuanced snapshot of a critical moment in time, in which the transition toward EVs, and thus the high-stakes obsession with Bolivia's lithium reserves, accelerated with a speed that was both undeniable and utterly confused. And to what end? We might ask the same about the global energy transition as Gogol asked of the rushing troika in *Dead Souls*: Whither, then, are you speeding? Whither? Answer me! But no answer comes.

FROM LITHIUM DREAMS TO LITHIUM GEOPOLITICS

When considering the question of a country's lithium reserves, size very much does matter. It wasn't always the case. When I began research on Bolivia's lithium project in 2019, the generally agreed-upon problem was not supply—potential or otherwise—but something like unrealistic demand. The discourse among lithium market influencers, investors, producers, and national and international energy policymakers, among other actors in the vast global lithium-energy assemblage, was that either there was already enough lithium in circulation to meet existing needs or that well-established lithium producers in places like Chile and Australia would be able to ramp up production to handle the growth in demand based on even the most ambitious predictions for the transition to EVs.

In light of this consensus around existing and future lithium supply, the question of Bolivia's lithium reserves was treated in a particular way: although the significance of the country's lithium deposits was never in doubt, especially as far as the Bolivian government, which had been trying to establish a lithium industry for years, was concerned, beyond Bolivia, the operative word was "instability." Because of the country's reputation for political and social unrest and their presumed lack of economic capacity, a wide range of institutions and market analyses dismissed the viability of lithium production in Bolivia as a tantalizing idea that was not worth the financial or political risks. As far back as 2010, the journalist Lawrence Wright had examined this framing of lithium potentiality in Bolivia in a *New Yorker* piece he entitled "Lithium Dreams."[1] He poses the question, "Can Bolivia become the Saudi Arabia of the electric-car era?" and then goes on to answer largely in the negative.

His trenchant assessment captures the prevailing sentiment at the time about Bolivia's lithium ambitions. As Wright puts it, the country's Movement to Socialism (MAS) government was "prone to revolutionary declarations" like "either capitalism dies or else Planet Earth dies," even though such rhetoric "tends to scare away the kind of foreign investment" that was a precondition for lithium industrialization in the country. This rhetoric exacerbated the fact that, according to Wright, the region of the Salar de Uyuni was marked by a kind of otherworldly lack of basic infrastructure. As he puts it, in a phrase that invokes an unvarnished colonialist imaginary, "Before Bolivia can hope to exploit a twenty-first-century fuel, it must first develop the rudiments of a twentieth-century economy." And because the global market for lithium could live without Bolivia's reserves, there was little incentive for governments or investors to help Bolivia build these supposedly missing rudiments.

But beginning in late 2019, the status of Bolivia's lithium reserves shifted, slowly at first, then rapidly. The trajectory of this dramatic shift underscores the ways in which lithium production is shaped in equal measure by institutional, ideological, economic, and technological forces. The fifth "synthesis report" of the Intergovernmental Panel on Climate Change (IPCC), the principal international authority on the human-induced climate crisis, was released in 2014. This sobering scientific reaffirmation of the planetary perils of climate change and, more broadly, the scope of catastrophic anthropogenic transformation, formed the basis for the landmark 2015 Paris Accords, during which 195 countries signed a treaty that promised, among other things, to limit the rise in global temperature to 1.5 degrees Celsius above preindustrial levels and to work collectively and globally toward net zero carbon emissions by 2050.

Following the 2015 Paris Accords, the IPCC was charged with preparing a follow-up report that would provide a more concrete strategy for how the world could meet these targets. In 2018, the IPCC published a "special report," which asserts that while meeting the 1.5 degree and 2050 net zero targets is at least scientifically possible, doing so requires revolutionary changes in global energy use, land use, and economic production.[2] In particular, the IPCC report focuses on the absolute necessity of replacing fuel-burning transportation with EVs, ideally charged by local, national, and interregional green electricity grids. Yet like so many international agreements, both the Paris Accords and the stern recommendations of the 2018 IPCC special report were left to the vagaries of Westphalian sovereignty on the one hand (for example, the United States left the agreement in 2020, only to rejoin it a year later) and, on the other, to the capriciousness of global capitalism, in which the question of the energy transition—including EV adoption—is shaped by logics that are quite different from those of climate change science.

But in early 2020, this sociopolitical and political economic calculus was upended—or, at least, disturbed—when the European Commission approved the €1 trillion European Green Deal (EGD), which promises to make Europe the world's first carbon neutral continent by 2050. To accomplish this, the EGD requires sweeping changes in transportation within much shorter timeframes, including a 55 percent reduction in emissions from cars by 2030, and, even more demanding, zero emissions from new cars by 2035.[3] Significantly, the EGD frames the transition to EVs as a signal marker of a "third industrial revolution," in which carbon neutrality and environmental protection are perfectly consistent with national and regional economic growth, resource (albeit "sustainable") extraction, and high levels of employment that are not "undermined by unfair competition from abroad."

Given the fact that the EGD represents a major political and economic intervention, one that impacts twenty-seven countries, 450 million people, and the world's third largest economy, its ramifying consequences for EV policy; lithium-ion battery production; lithium supply; and, eventually, the status of Bolivia's lithium reserves were undeniable. In the months after the launch of the EGD, and despite the upheaval of the global Covid-19 pandemic, most the world's major automakers reacted, announcing their own corresponding transitions to all-EV or mostly EV production lines within a variety of time horizons: 2030, 2035, 2040, and so on. The reason this is so important is that the shift from fuel-burning cars and trucks to EVs by the producers of such vehicles is where the rubber meets the road. Instead of the grand gesture of international agreements and political posturing, such a shift demands wholesale and structural changes, many of which take years to realize, including the creation of entirely new supply chains, the remaking of industrial facilities, and the reshaping of labor forces.

At the same time, the planned EV transition at the level of production and distribution was met with the proliferation of city-level versions of the EGD, a strategy compelled by the EGD itself. Major cities throughout the EU, and some within Europe but outside the union (like Lausanne, Switzerland, my current hometown), adopted their own green deals or climate plans, which imposed even stricter controls on transportation within even tighter time horizons (for example, bans on fuel-burning vehicles within city limits by 2030). All of these measures, taken together, meant that European EV and climate policy became the catalyst for a more global transformation in EV production and policymaking, a transformation that could be seen in everything from the notable increase in EV lines by Chinese automakers, who hoped to break into the European and other major markets, to the 2022 Inflation Reduction Act in the United States, which included $12 billion in incentives to encourage EV adoption by American drivers (mostly through tax credits).

And with these major changes, the prevailing discourse around lithium supply and demand underwent its own transformation. If, as late as 2019, most lithium analysts were arguing that the existing supply of battery-grade lithium carbonate was more than sufficient to meet predicted demand, by the end of 2020, the consensus perspective had been turned on its head. With estimates for demand now rising to tens, and eventually hundreds, of millions of new EVs, the question suddenly became whether or not enough lithium-ion batteries could be produced to meet such soaring growth. And although lithium-ion batteries are composed—depending on the technology and chemistry—of a number of so-called critical minerals, including nickel and cobalt, each with its own fraught resource history, lithium is by far the most important.

This means that the supply of lithium emerged as both the main limiting factor for one of the key pillars of global climate change policy—the electrification of mobility—and the basis for a new energy geopolitics, a brutal competition over yet another nonrenewable resource. With demand estimates far outstripping global supply, which even factored in dozens of *planned* extraction projects (both largescale and relatively minor), the struggle over lithium began to resemble the epochal transition that took place when coal power was supplanted by oil.[4] Now, with both green energy policymaking and the widespread shift toward EV manufacturing as signs on the wall, it was the Age of Oil itself that was giving way to an era in which lithium-powered forms of energy might eventually replace hydrocarbons at the center of a new global resource politics and power nexus.[5]

Without wanting to give undue—and, perhaps, unanthropological—importance to price fluctuations, two datapoints seem particularly relevant to demonstrating how a global obsession with lithium supply, inflected by growing desperation, was reflected in rapid changes across the wider lithium value chain. At the end of 2019, the average price of battery-grade lithium carbonate hovered around $8,000 per metric ton. By the end of 2022, the price had skyrocketed to around $75,000 per metric ton.[6] For the same twenty thousand tons of battery-grade lithium carbonate, for example, this is the difference between $160 million and $1.5 billion in revenue. And what about Tesla, the world's largest EV company? At the end of 2019, its "market cap"—the price per share multiplied by the number of outstanding shares—was about $76 billion. At the end of 2020, less than one full year after the EGD was announced, Tesla's market cap had spiked to almost $700 billion. And by November 2021, at its highpoint, Tesla was worth $1.3 trillion, making *an EV company* more valuable than any oil company in the world except Saudi Aramco and five times more valuable than Toyota, the world's second largest car company. Indeed, during this period, Tesla's value was greater than that of the next nine largest car companies *combined*.

Which brings me back to the question of the size of national lithium reserves. For Bolivian geologists, this is a question with political, technological, and reputational—as much as economic—implications. In 2020, Guido Quezada Cortez and his colleague Nelson Carvajal Velasco published a detailed summary of a five-year study of lithium undertaken at the Salar de Uyuni.[7] As will be seen later in the book, this was also the year in which Quezada had been fired as one of the chief geologists for YLB during a time of political turmoil. But between 2013 and 2018, he was part of the team that conducted an extensive study of the lithium concentrations in the salar's brine. Quezada was passionate about promoting and even defending the results of this study.

On the one hand, he emphasized the extent to which his team of Bolivian geologists and hydrographers had the capacity to arrive at a highly accurate estimate of the lithium reserves in the salar—and they did so in a way that showed the technical limitations of the existing estimates. The latter were based primarily on surveys done by the French geochemist François Risacher working on behalf of the French government's Office of Scientific and Technical Research Overseas (ORSTOM), an institution originally founded to oversee "colonial scientific research" during the wartime Vichy regime.[8] On the other hand, Quezada was concerned that the new estimates, which radically altered both the scientific and discursive framing of Bolivia's lithium reserves, would be received with confidence by both an international audience and, more importantly, Bolivians themselves, who were being asked by the MAS government to view the lithium project as the centerpiece of the country's future.

When Quezada and the Bolivian research team began their reanalysis of the salar's brine in 2013, the accepted figure was 8.9 million tons of lithium reserves. But when YLB released the results of the five-year study in 2018, in a document meant largely for governmental use, Quezada and his fellow scientists had made a blockbuster discovery: there were actually more than *twenty-one million* tons of lithium in the salar. Moreover, as Quezada and Carvajal explain in their 2020 analysis of this landmark study of lithium in Bolivia, the estimate was based only on sample perforations of the Salar de Uyuni. As I would later come to learn, preliminary research on Bolivia's two other large salares, the Salar de Coipasa and Pastos Grandes, that used the same techniques Quezada and his team had developed for the Salar de Uyuni, indicated that there could be up to ten million *additional* tons of lithium reserves, bringing the more likely estimate for Bolivia's three major salares to thirty-one million tons.

The stunning announcement from Bolivia about its lithium reserves, reinforced by the technical detail and comprehensiveness of the historic five-year investigation, was adopted by the U.S. Geological Survey in its widely used ranking of countries. In 2020, just at the global tipping point moment for lithium, the USGS released its yearly report: among the twenty-three countries with appreciable reserves of lithium, a total that was almost eighty million tons worldwide, Bolivia had suddenly shot to the top, with a quarter of the world's reserves.[9] And when the likely lithium in the brines of Coipasa and Pastos Grandes were included, Bolivia held almost 40 percent of the world's known reserves based on the 2020 figures. To put this in perspective, Bolivia's lithium reserves were more than Argentina's (#2) and Chile's (#3) combined, or thirteen million tons more than the combined reserves of the United States (#4), Australia (#5), and China (#6). Again, from the perspective of global energy policy and Bolivian policymakers alike, so much depends on Bolivia's lithium.[10]

EXTRACTING THE ENERGY TRANSITION, FLEXIBLY

But as will be seen in more depth in the next chapter, Bolivia has been in this position before, many times before. To different degrees of global magnitude and with different consequences for Bolivia's economy, political history, and conflicts over territory, its natural resources have for centuries been at the center of large-scale transitions, all of which have involved an often toxic combination of competition for something considered scarce, the use of various forms of violence (economic, political, military, symbolic), new technologies, and the essential linkage between extraction and disparate visions of the future. This perspective from the *longue durée* is important when considering the ways in which Bolivia's lithium project is entangled with the global energy transition. It suggests that there is little that is unproblematically unique about the race to unlock the planetary potential of Bolivia's lithium reserves, despite the wider framing that invests the lithium-extraction-to-EV nexus with a kind of moral urgency that itself comes from a pervasive existential panic that is so characteristic of the high Anthropocene.[11]

Nevertheless, even if there is value in placing the struggle over lithium within the context of longer histories of extractivism, it is also worthwhile reflecting on what the importance of lithium says about *this* particular transition and its various political claims, material realities, and moral posturing. What are the wider implications of the fact that the energy transition, itself compelled by both political mandate and transformations in productive technologies, is fundamentally dependent on the extraction, processing, and commercialization of a nonrenewable resource? Or, to put this another way, *if* the green energy transition is so reliant on, indeed, impossible to imagine without, the extraction of millions of tons of lithium and the manufacture and distribution of hundreds of millions of EVs, what does this reveal about the green energy transition itself?

In using the ethnographic, historical, and critical study of lithium both in Bolivia and beyond as the basis for responding to these questions, this book has been nourished by the research and analyses of many other colleagues working at the crossroads where lithium meets energy policy, climate change mitigation, and socioeconomic dislocation. Not surprisingly, much of the literature on lithium concerns developments in Chile,[12] Argentina,[13] and Bolivia, including important works by Bolivian scholars and environmental activists.[14] Other scholars have conducted comparative studies of lithium processes and imaginaries *across* these three countries.[15]

This book has also been nourished by the vast literature that examines the economic, productive, material, social, and historical dimensions of extractivism.[16] Indeed, partly as a critique of the centrality of this literature, a group of scholars has recently argued that the concept of extractivism itself has reached its limits, both theoretically and as a category for descriptive analysis.[17] Yet in situating the study of lithium in Bolivia in relation to the voluminous existing body of work on extractivism—including, again, important studies conducted by Bolivian scholars—I find it useful to distinguish between two senses of extractivism, both of which are used as framing devices at key junctures in the book.[18]

First, it captures a particular field of capitalist production at the points at which the raw materials of wider economies—the political economic building blocks—are accessed, refined, and processed: in short, converted into the first forms of capital. As a mode of production, extractivism is associated with a spectrum of violent consequences, from environmental despoliation to the suppression of anti-extractivist resistance (often mounted by historically marginalized communities) to the impoverishment of countries through what is often a compelled economic reliance on one or a handful of primary commodities.

Second, extractivism usefully describes the ways in which the emergence and organization of a raw-material-dependent mode of production takes place as part of wider histories of political conflict and economic plunder. Extractivism can thus also be understood as the *political* logic of various processes of global and regional (neo)colonialism through which demand for raw materials is both a prerequisite and justification for intervention, the extension of control (both direct and indirect) over states and populations, and the cultivation of different kinds of dependency.

At the same time, as a pernicious resource geopolitics of planetary scope, extractivism often implicates national, regional, and local elites within countries enmeshed in the extractivist trap. This fact can complicate efforts to easily map the topographies of extractivism in a particular place and time or to chart what Penelope Anthias has described as the "counter-topographies" of extractivism—that is, instances of resilience and community mobilization within extractivist ruins.[19]

Moreover, it is essential to resist the tendency to think of extractivism-as-resource geopolitics in overly dichotomous terms despite the fact that identifying global patterns of extractivism can be useful for heuristic or comparative purposes. For example, in her research on social conflict over a proposed lithium mine in Serbia, Nina Djukanović shows how what she calls "extractivist frontiers" are established in ways and in places that defy overly simplistic understandings of how contemporary resource boundaries are drawn and contested.[20]

Extractivism as both mode of production and resource geopolitics becomes "neo-extractivism" under a number of conditions that are relevant to the study of Bolivia's lithium project and its relation to the global green-energy transition. Debates around new forms of extractivism have been shaped in particular by South American scholars like Eduardo Gudynas[21] and Maristella Svampa,[22] who have analyzed the differences and similarities between older extractivist histories and new projects of state-led developmentalism in South America, in which the state asserts control over the production of raw materials as part of national strategies of social change and economic redistribution. As with extractivism, scholars—within South America or otherwise—deploy the concept of "neo-extractivism" as a sharp tool of critique. This critique takes on an even sharper edge because the charge of neo-extractivism is also meant to call out the hypocrisies and contradictions of national development plans that continue to rely on environmentally and socially harmful practices while asserting dubiously radical shifts in the status of resource sovereignty.

Finally, the more general category of extractivism—again, as both mode of production and as resource geopolitics—has been adapted to studies of different aspects of the green energy transition. Given that various forms of supposedly green or sustainable energy are dependent, to varying degrees, on the extraction and processing of raw materials, the question becomes whether these processes can and should be analyzed through the wider extractivist critique. For scholars and activists who make this linkage, the charge of "green" extractivism is a way of pushing back against the epochal pretensions of the broader energy transition along political-economic, environmental, and sociopolitical lines, among others.

For example, in a comparative study of resource governance across the three countries of the Lithium Triangle (Chile, Argentina, and Bolivia), Daniel Voskoboynik and Diego Andreucci argue that discourses of green economic development and climate-friendly or sustainable mining are undermined when they seek to justify the ongoing need for mineral-intensive extraction.[23] And in research on conflicts fueled by the eagerness of the European Union to support the development of what would be Europe's largest open-pit lithium mine in northern Portugal, a project that is being overseen by the London-based private company Savannah Resources, Alexander Dunlap and Mariana Riquito contend that green extractivism is an "insidious social technology," one that uses a generalized language of climate justice to obscure the realities of anti-mining activism, local political resistance, and environmental harm.[24] Scholars have even extended the green extractivist critique to the analysis of less mineral intensive technologies that might not rely to the same degree on the extraction of "critical" minerals but that are nevertheless embedded in similar supply chains and political economies. Even the elaboration of wind power projects can depend heavily on strategies of state control and the suppression of community opposition, something that Cymene Howe and Dominic Boyer have described as "aeolian extractivism."[25]

But if both extractivism and neo-extractivism are necessary—albeit by now well-worn—frames for making broader and comparative sense of Bolivia's lithium project, they are not sufficient. This is because, as will be seen most directly in chapter 3, Bolivia's approach to lithium industrialization underwent a subtle, if profound, shift, one in which the Bolivian state reformulated its orientation toward lithium in response to a cascade of *resistance*: political, economic (both macro- and micro-), environmental, technological, and material. Through this reformulation, Bolivian policymakers increasingly stopped trying to square the neo-extractivist circle in which commitments to social and economic change and environmental protection clash with the need to extract nonrenewable resources as an imperative of national development.

Instead, the ideological grounding of the lithium project began to loosen, to become more flexible, in ways that defied both the existing political economic model and the critical pigeonhole of neo-extractivism. In an echo of David Harvey's argument about the ways in which late twentieth century capitalism developed more flexible approaches to accumulation, I think something similar can be said about Bolivia's evolving lithium strategy:[26] just at the moment in which resistance at many levels threatened to undermine the lithium project, with perhaps fatal consequences, the government pivoted toward what might be called *flexible extractivism.*

This loosening and recasting could be seen in everything from industrial policymaking to YLB's relations with social movements in the salar region. Especially after the election of Luis Arce in 2020, the lithium project became infused with a deep ideological hybridity and a thoroughgoing political and technological pragmatism. Although a rhetoric of environmental steward-ship and gestures toward Indigenous eco-ontologies (like *vivir bien*, or living well) remained in various official documents, the actual practice of lithium industrialization was marked by a focus on production at all costs and a will-ingness to reimagine the relationship between extraction and sovereignty. And because lithium extraction, unlike oil and gas, is the basis for a technology—lithium-ion batteries—that is a pillar of a supposedly more sustainable future, Bolivian policymakers seemed to be making up this new ideological terrain even as they walked on it.

More concretely, the shift toward flexible extractivism had a number of implications for the unfolding lithium project. First, at the level of infra-structure, it opened the door to a much wider range of possible international partners, including private companies from the United States, something that would have been unthinkable only a few years before. Second, it allowed the Bolivian government and the state lithium company to finally acknowl-edge the underlying material, technical, and economic limits to the grand ambitions of the lithium project's early years, including the ambition to com-plete the lithium value chain from the production of battery-grade lithium carbonate to the manufacture of EVs at a commercial scale. And finally, the turn toward flexible extractivism introduced an ideological ambiguity into the lithium project that obscured actual and potential conflicts around dif-ferent visions of resource extraction.[27] Indeed, the emergence of flexible extractivism as the final stage of neo-extractivism—at least in the context of Bolivia's lithium project—reoriented the field of social and political struggle through a kind of ideological withdrawal that was replaced by a lower-grade focus on technical capacity, industrial efficiency, and debt repayment.

SCALES OF ENERGY, IDEOLOGY, AND ANTHROPOLOGY

Having situated the study of Bolivia's lithium project in relation to a number of broader contexts, let me now turn to the underlying research, including its disciplinary and interdisciplinary anchorages, in more detail. As a long-term anthropologist of Bolivia, it was perhaps inevitable that I would want to learn more about the nuances of the country's lithium project. While completing follow-up research for another book,[28] the topic of lithium was raised repeatedly by interlocutors in Bolivia, at the same time that the country faced increasingly extreme effects of climate change, like megafires.[29] Fortunately, I received funding to direct an international study of the lithium project just at the moment at which interest in Bolivia's lithium reserves was intensifying, for the reasons I have explored above.

The research unfolded within a framework that was both thoroughly anthropological and interdisciplinary. The questions I pursued and the wider theoretical starting points were shaped by debates and insights from the vibrant anthropologies of energy, climate justice, and resource politics, while the research itself was informed by insights from different "partners" (as the funding agency describes them), including those with backgrounds in geology, chemical engineering, marketing, electrical engineering, business administration, and tourism, among others. And these were just the official project members; in addition, as will be seen in different ways throughout the book, the ethnographic data that my research produced over four years were fundamentally shaped by the perspectives and lived experiences of many people from a wide range of backgrounds and social positions.[30]

It is important to acknowledge that my own research tracked Bolivia's lithium project across key sites (and scales, see below) throughout and beyond the country rather than focusing on particular villages or social movements in the Salar de Uyuni region, although, as the opening to this introduction shows, I did conduct extended fieldwork in the towns of San Cristóbal and Colcha "K," research that is analyzed elsewhere.[31] But in the context of the wider research project, a division of labor was adopted in which I would follow the trail of lithium wherever it led while another member of the research team, David Schröter, would examine the impact of the lithium project on several communities nearest the sites of extraction.[32]

As a consequence, my research examined a range of institutions, positionalities, and lithium imaginaries. As will be seen throughout the chapters, this range included everything from frontline workers at the Salar de Uyuni to energy and lithium policymakers in La Paz to green energy entrepreneurs, both in Bolivia and in Germany and the United States. Some of this research followed the tradition in anthropology of "studying up"—that is, turning a critical ethnographic lens toward people and institutions in positions of relative power in order to pierce through official discourses, but also as a potential tool of social and political change.[33] Even more, the radiating optic of studying up has become increasingly necessary for research on the multiple entanglements that constitute the energy transition, especially given its imbrication within a global capitalist architecture that continues to perpetuate the kinds of Anthropocenic harms that the energy transition is meant to mitigate.

Nevertheless, given the fact that it soon became clear that Bolivia's lithium project was constituted by a vast heterogeneity of nodes, from the political to the technological, nodes that were also located in time and space, including spaces beyond Bolivia itself, I confronted early difficulties in settling on conceptual and methodological frameworks that seemed to be adequate to the considerable task. Here, too, I was fortunate in that the relevant literatures offered a number of inspiring guides, which I adapted and extended in relation to the research project's various complexities. The first and perhaps most important problem was to identify a methodological approach that would prove capable of steering and structuring research under difficult conditions: political (the November 2019—October 2020 coup d'état), environmental, and epidemiological (the 2020–2023 global Covid-19 pandemic). At the same time, this approach needed to point clearly toward an eventual theoretical framework, one in which our analysis of Bolivia's lithium project would speak meaningfully to both the project's ethnographic specificities and the ways in which these articulated with wider—regional, hemispheric, global—sites and processes within a loosely connected global assemblage.

Several potential models were available, and in the end, our research methodology could be said to incorporate elements of each. For example, because I conducted ethnographic research in what would, without complication, be understood as different distinct locations—the offices of YLB, the industrial facilities at the salar, a battery research and development laboratory, the headquarters of different green energy start-ups, and so on—I used a version of multisited research that has played an important role in anthropology and allied disciplines at least since George Marcus suggested that such an approach was essential for what he described as "ethnography in/of the world system."[34]

In addition, I was also influenced by Sally Engle Merry's nuanced extension of the multisited approach, something she described as "deterritorialized ethnography."[35] In elaborating on what it meant to study "placeless phenomena in a place"—in her case, the confounding placelessness of the

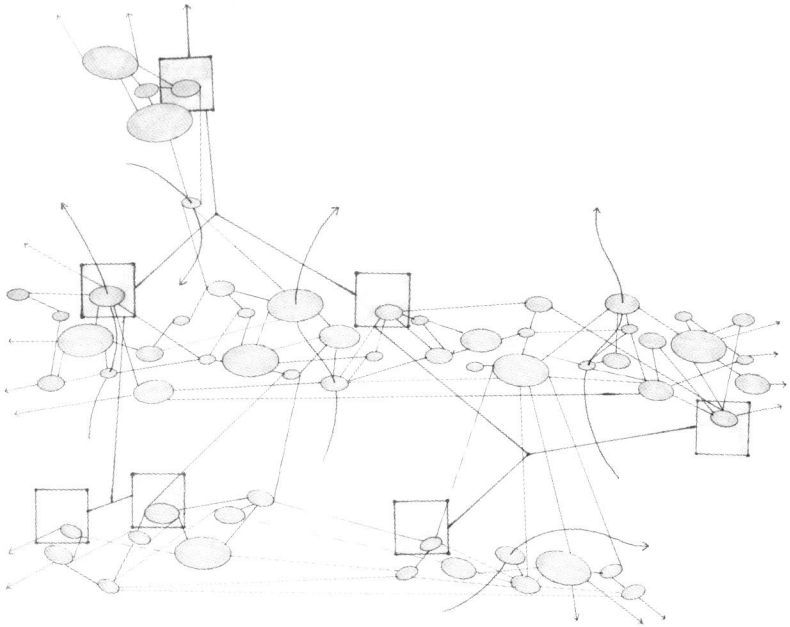

FIGURE 2. A global assemblage, as imagined by the Guyanese-Canadian illustrator Marc Ngui. Used by permission.

globalization of women's rights—Merry argues that the essential challenge is to "find small interstices in global processes in which critical decisions are made, to track the information flows that constitute global discourses, and to mark the points at which competing discourses intersect in the myriad links between global and local conceptions and institutions."[36] This is an approach ideally calibrated for research with the "fragments of a larger system that recognizes that the system is neither coherent nor fully graspable." To the extent to which Bolivia's lithium project is entangled with the wider energy transition—entangled through emerging productive alliances, resource policymaking, *and* discourses around a postcarbon future—it, too, constitutes a fragment of a wider "system" that is also equally difficult to fully apprehend, at least through the methods of social science research.

Yet perhaps the most important addition to the research project's methodology was one that allowed me to shine an analytical light on what is likely a phenomenon with wider relevance beyond the lithium energy assemblage. As scholars like Gabrielle Hecht have argued one of the most pressing challenges for social scientists of the Anthropocene is to reveal the critical particularities that mark widely diverse lived experiences of climate crisis, anxiety about the future, and what I have described elsewhere as conditions of foreclosure,[37] while at the same time keeping "the planet and all of its humans in the same conceptual frame."[38]

To do so, Hecht proposes a variation on multiscalar research that manages to navigate between the "scalar claims of historical actors and projects," including, importantly, the researcher's own scalar claims—in my case, about Bolivia's lithium project and its relation to the wider, if not exactly planetary, energy transition. Moreover, the concept of *scale* has the potential to do much more expansive work than either sites or deterritorialized discourses. As she explains, "Scale is not just about size or granularity. It is also about categories: what they reveal or hide, the ways in which they do (or do not) nest. And it is about orientation: how we position ourselves, what we position ourselves against, and what comparisons such locations do (or do not) authorize."[39]

In following this "interscalar" approach to the lithium energy assemblage, I came to realize that there was an essential structural *décalage*, or unbridgeable gap, between this fundamental "category" of climate change mitigation and the actual planetary processes in relation to which the electrification of mobility is perceived to be one among several decisive responses. If the lithium-extraction-to-EV nexus is actually a fragmentary assemblage comprised of shifting scales of production, history, practice, and ideology, among others, this description contrasts sharply with the ontologies of the Anthropocene itself, including those portrayed by climate science, ecology, and paleontology. Although one *must* speak of—one must "authorize"—the ways in which the Anthropocene is defined by catastrophic changes to various Earth *systems* understood as tightly interconnected (if complex) wholes, something like the opposite is true of the assemblages that are supposed to be the main bulwarks against these changes.

If the assemblage whose diffuse contours are examined in this book is similar to the others at the center of the energy transition, whether organized around solar power, wind power, or other less carbon-intensive or even carbon-neutral technologies, then what can be said is that they are indeed characterized by a loose coherence, yet a coherence that is far outweighed by ad hoc interconnections, situated articulations, and even a kind of randomness that can only be explained—for example, through ethnographic research—but not generalized, especially for purposes of policymaking. The fact of this décalage has profound implications for the future of climate change mitigation. In short, there is a basic incompatibility between the planetary and ecosystemic dimensions of Anthropocenic harm (across the full range) and our capacity to design "systems"—including energy systems especially—to counteract, if not reduce, these harms.

This realization also changes how critical social science research, including anthropological research, on the energy transition must be understood. Rather than producing engaged studies that are ultimately directed toward advocacy for "green" energy or supposedly more sustainable modes of living, which can be of much value on their own terms (but not in Anthropocenic terms), I would argue that we must view social research now as a project of revealing the ways in which people and communities find meaning amidst what are the "remains of the future, the ruins of chimerical global visions that will never come to pass."[40]

CONCLUSION: LITHIUM IMAGINARIES AND LIVABLE COLLABORATIONS

In many ways, Bolivia's lithium project—despite its ethnographic singularities—epitomizes so much of what is paradoxical, irresolvable, and ultimately bedeviling not simply about the unfolding energy transition but about our widely diverse plight in the face of worsening Anthropocenic deterioration. Yet even if several of these broader points of illumination underscore the daunting challenges and potential for continuing violence—"slow"[41] or otherwise—as the energy transition picks up pace, the study of lithium also reveals forms of social adaptation and solidarity, as well as a willingness to resist economic and political imperatives that do not reflect the fact that responsibility for planetary harms such as climate change is itself not planetary. As with other histories of injustice, the climate crisis is a powerful reminder of how the many usually suffer because of the predations of the few.

If these mixed implications of the current study can be divided into the critical (rather than the negative) and the inspiring (rather than the hopeful), let me begin with the critical. As will be seen throughout the book, the category of technology takes on an importance that pushes aside many other considerations—environmental, social, political, and ethical. From extractive technologies to debates over battery cathodes to EV design, the lithium energy assemblage is fundamentally shaped by what might be thought of as technological regimes. But as the book demonstrates, the privileging of technology, the massively expensive and exclusionary political, economic, and social bet on the technological fix to Anthropocenic harm, represents one of the greatest barriers to truly undoing or even ameliorating this harm, let alone achieving anything approximating climate justice.

At the same time, the study of Bolivia's lithium project illustrates the ways in which the assemblages implicated in the energy transition are deeply interwoven into the political economic framework of global capitalism, regardless of whether (as in Bolivia) national economies are formally committed to socialist principles in planning, control, or the use of revenues. Indeed, as the power of Chinese state companies across the entire lithium energy assemblage shows, the logics of green energy capitalism are entirely compatible with the institutional and political structures of state socialism. Yet whether shaped by public or state-owned enterprises or those anchored in the more conventional private sector, the green energy transition bears all of the warping marks of capitalist penetration, which depends on global market forces that have little concern for wider planetary suffering.

And to return to the concept of extractivism—as both mode of production and resource geopolitics—the ethnography of key nodes in the wider lithium energy assemblage drives home the more fundamental argument that extraction will never form the basis for a real transition to a truly sustainable, equitable, and just future. From the exploitation of critical minerals to the development of vast new energy infrastructures to the mobilization of "insidious" discourses of green capitalism, the importance of extraction to these processes signals their actual nature, what interests they can and do serve, and the ways in which the very structure of the energy transition must be understood as one of the key impediments to meaningful change of our time.

Even so, the study of Bolivia's lithium project and its multiscalar entanglements with the wider energy transition also carries lessons of inspiration and possibility. In examining the ways in which quite diverse imaginaries emerge around lithium and its role at the heart of climate change mitigation and energy policy, this book shows how the capacity to envision, project into the future, and confront the past endures even under conditions of political, economic, and ideological compulsion. To imagine lithium otherwise in the midst of—and even against—the energy transition is itself a potentially transformative act.

Hints of transformation can also be seen across the many different forms of resistance that take place within the energy transition—resistance not just to extraction and the various havoc it wreaks but to the illusory futures promised by the energy transition. Even more, resistance to the juggernaut of broader resource geopolitics, including those in which Bolivia's lithium project is ensnared, is an important reminder that Indigenous and peasant communities have faced the full brunt of Anthropocenic harm for centuries, from the beginnings of mercantile colonialism to the climate crisis of the present.[42] In keeping the critical focus as much directed toward this longer past as to the impossible future promised by new energy technologies, these communities offer a glimpse into what an alternative—and compelling—account of climate justice might look like.

And finally, the study of lithium in an era of energy transition—its material complexities, its fitting near-weightlessness, the ways in which it infuses different (and often contradictory) forms of desire, even its hidden coursing through landscapes that are as forbidding to those who live around them as to those who come to exploit them—opens an often unexpected window onto the possibilities for what Anna Tsing describes as "livable collaborations, even against a background of massive human destruction.[43] In the face of the stark contradiction of a green energy transition grounded in the destructive extraction of nonrenewable resources like lithium, people nevertheless manage to think together and beyond this contradiction; to build their own sustainable worlds, however limited in time and space; and to forge their own localized paths toward more modest futures of their own making.

ONE

─────────

Tracing the Prelives of Lithium

FOR MOST OF MY LIFE, like many people I imagine, I hadn't given much thought to lithium. Even during the many periods of ethnographic fieldwork in Bolivia over the years, including during the different Movement to Socialism (MAS) governments of Evo Morales (2006–2019), the question of lithium never came up in the midst of much more pressing conflicts over the new constitution; anti-Indigenous violence in the country's eastern departments; or, somewhat later, plans by the Morales administration to extend its political power through a series of contested stratagems. Of course, as I would come to learn, lithium *was* a critical issue in Bolivia and beyond throughout this period, if one knew where to look or the right questions to ask. But circa, say, 2014, over morning coffee and an extended perusal of the newspapers from my base in La Paz, I would have likely seen articles in the economics section of *La Razón* that described one development or another in the country's slowly unfolding lithium project. Yet this wouldn't have drawn my attention, especially since during these years the main focus of economic development was the state-controlled natural gas industry: new contracts with foreign buyers, the impact of price fluctuations, the growth of yearly revenues, critical reflections on how long the gas "bonanza" would last, and so on.

Looking back farther in time, without any specific reasons to work with, consume, or otherwise seriously consider lithium, perhaps the only resonance it had for me that I can readily recall is the 1991 Nirvana song "Lithium." But even here, tellingly, the reference is more ambiguous than it would seem. The lyrics themselves make no mention of lithium and Kurt Cobain described the song as a reflection on the way some people embrace religion—as he did during his teens—as a way of coping with life's traumas. The still-passionate

THE fuel propellant of the future may prove to be an inorganic material, with metallic properties, capable of releasing tremendous heat burning capacity. Such an inorganometallic will likely contain a compound of *lithium*. For *lithium* offers uniquely valuable properties . . . properties that aid in contributing an unusually high power-to-weight ratio so necessary for military missiles and rockets.

Lithium, for example, combines low density with high heat of combustion to give a much sought after ratio of extraordinary chemical energy per unit of weight. On this score alone it proves of inestimable value.

Will these properties improve your product?

. . . low density	. . . high flash point
. . . high heat capacity	. . . easily cut with a knife
. . . high heat of fusion	. . . ductile, can be extruded and rolled
. . . chemically reactive	. . . readily melted or cast
. . . low melting point	. . . lighter than magnesium or aluminum
	. . . can be dispersed in suitable media

Consult our PR&D department on your use-research problems. Up-to-date Product Data Sheets plus laboratory quantities of lithium metal, metal dispersions, metal derivatives and salts are yours for the asking.

. . . *trends ahead in industrial applications for lithium* LITHIUM CORPORATION
OF AMERICA, INC.
2685 RAND TOWER, MINNEAPOLIS 2, MINN.

PROCESSORS OF LITHIUM METAL • METAL DISPERSIONS • METAL DERIVA-
TIVES: Amide • Hydride • Nitride • SALTS: Bromide • Carbonate • Chloride •
Hydroxide • SPECIAL COMPOUNDS: Aluminate • Borate • Borosilicate • Cobaltite
• Manganite • Molybdate • Silicate • Titanate • Zirconate • Zirconium Silicate

BRANCH SALES OFFICES: New York • Pittsburgh • Chicago • MINES: Keystone,
Custer, Hill City, South Dakota • Bessemer City, North Carolina • Cat Lake,
Manitoba • Amos Area, Quebec • PLANTS: St. Louis Park, Minnesota • Bessemer
City, North Carolina • RESEARCH LABORATORY: St. Louis Park, Minn.

FIGURE 3. Advertisement from the Lithium Corporation of America, 1957.

annals (and discussion websites) of Nirvaniana are filled with speculation about Cobain and his possible use of lithium to treat a case of bipolar disorder, but since everyone agrees that he was never formally diagnosed with the disease, he wouldn't have had access to it through a medical prescription. His drug of choice, unfortunately, was heroin, and so perhaps his use of "lithium" for the song's title was more random, something that came from a free association between religion, mental illness, and medication.

But my understandable indifference toward lithium came to a sudden end in 2017, in the months after the Bolivian state lithium company (Yacimientos de Litio Bolivianos) was founded. Discussions about lithium extraction became more common in national public debates, which meant, among other things, that the subject of lithium became politicized in ways both uniquely Bolivian and common to struggles in other countries over resources and national economic development. Even more, *litio* developed into an intriguing buzzword that appeared across a growing range of contexts, from presidential or vice-presidential speeches about the country's commitment to environmental conservation to academic workshops on the possibilities for economic growth after the end of what Bret Gustafson has rightly described as Bolivia's "Age of Gas."[1]

Within months, my natural disregard for lithium was replaced by a keen interest tempered by something like unease: if this mineral or element was becoming critical not just for Bolivia's economy but for the wider global energy transition, how was I to ever really understand it? As an embarrassingly simple starting point, what *was* lithium anyway? Yet very soon, this unease gave way to a deep fascination; as will be seen below, lithium has been seducing different people for different reasons for well over a century. This chapter is in part a study of this history of seduction, of allure, of promise, of a curious linkage between industrial chemistry and futures transformed.

My own changed relationship with lithium has been—at least for me—unusually embodied. It is one thing as an anthropologist to spend years and even decades conducting research on social and political movements, contrasting ideologies, law and society, and the practice of human rights, all of which are expressed in concrete ways and yet remain abstract, theoretical, discursive. It is quite another to take the lightest metal in the universe (atomic number 3, just after helium) seriously as an object of anthropological study; indeed, to place it at the very center. So much of what has made lithium variously irreplaceable over a long series of applications takes place at levels that are invisible to the human eye, anthropological or otherwise. But this unavoidable fact doesn't lessen the tangibility of lithium. In many ways, the elusiveness of lithium at human scales only heightens the importance of its multiple expressed materialities—historical, chemical, economic, and sociopolitical.

Standing on the Salar de Uyuni, the clash between the visual disorientation of utter flatness stretching beyond the horizon in all directions and the visceral knowledge that beneath this flatness a salty liquid holds millions of tons of the most sought-after alkali metal in the world couldn't be starker. As a phenomenological proposition, the contrast is almost impossible to accept. Even after years of careful ethnographic consideration and of grappling with the sociomateriality of lithium, with its chemical signatures and hydrogeological formidability, it remains for me an element with a peculiar, even frustrating, kind of agency, always flowing unseen and unapprehended just below one impenetrable surface or another.

Yet for some reason, when lithium's briny recalcitrance has been overcome and it has been captured in a medium that renders it usable, it becomes a substance that I have been repeatedly and surprisingly invited to ingest. This years-long attempt to convince me to sample the wares, to taste the goods, to test out the product, began in August 2019 during my first research visit to the YLB facilities at the salar. Augusts at the Salar de Uyuni present a slew of additional contrasts, unconnected to the implausible topography. Sparkling blue skies, magnified by the thin air at four thousand meters, provide a dawn to dusk space for the unencumbered passage of the tropical sun, which remains impressively overhead even during the southern winter. But despite the glowing sun, which bears down from a verticality that makes it seem closer to Earth, at least for someone raised in the northern latitudes, the days are usually bitterly cold. And there is no escape from the pervasive chill: even though the sun's radiation, coupled with the altiplano's elevation, is the driving force that turns lithium-rich brine into compounds on which the global energy transition depends, it seems to offer a kind of anti-warmth. The longer one spends basking in the August sun at the salar, the colder one gets.

It was on one of these frigid tropical days that I was first offered lithium, an experience that felt vaguely illicit. At the end of my first survey of the different operations, including the evaporation ponds and YLB's industrial potassium chloride plant, I was shown an older building—one of the first— where YLB scientists learned how to convert adulterated salts into "technical grade" (which is less pure than "battery grade") lithium carbonate. One of the YLB officials, someone attached to the company's public relations division, had accompanied me throughout the day. Despite what felt like a creeping hypothermia, which made taking notes increasingly impossible, I passed my first day at ground zero for Bolivia's lithium project in a state of wonder bordering on giddiness, something my chaperone had obviously noticed.

When we finally arrived at what was clearly meant to be the culmination of a carefully designed passage from evaporation ponds to the crowning glory of national chemical engineering, the extremely difficult production of lithium carbonate, he logically assumed that I would want to celebrate this achievement—by having a taste of the inert white powder. Reaching for a small unmarked bag, he opened it wide and held it out. *Try some*, he said, *you can eat it*. Not knowing how to respond but suddenly warmed by an adolescent feeling of peer pressure, I looked at him with worried doubt: does participant observation, an anthropological hallmark, require possibly poisoning myself with lithium on my first day of fieldwork?

Sensing my hesitation, my guide quickly said, *It's completely safe to eat, I think lithium is used as a medicine.* He then touched his pointer finger to his tongue, swirled it around generously in the bag, and licked it clean. He stared at me for a fleeting instant as if he was both processing what I took to be the powder's strong chemical aftertaste and, at the same time, reminding himself of what his YLB colleagues had told him about its relative harmlessness. Reassured, he offered me the bag for a second time. I refused again, which clearly disappointed him. In what now was certainly a gesture of both defiance and national pride, he swirled his finger in the bag once more, popped it in his mouth with relish, and then ended the tour with what was certainly less regard for me than when it began.

This first failed offering was followed by others over the years, each of which was met with the same refusal—polite, I like to think—on my part. Would I have come to understand lithium better, or in a different way, had I finally relented and agreed to send its molecules coursing through my bloodstream like the brine that courses through the salar's hidden strata? Maybe. But in the end, I take my own microhistory with lithium's alimentary potential as yet one more episode in a much longer and more consequential chronology, one whose outlines must be appreciated well before confronting the ethnography of Bolivia's unfolding lithium project and its relation to the wider energy transition.

Indeed, this chronology complicates the received approach to contextualizing the extraction of resources like lithium and the impact of its associated infrastructures. As Thomas Yarrow has shown, through an ethnography of an unrealized resettlement project in Ghana, the *afterlives* of failed or incomplete development projects can constitute their own kind of ruination, just as much—albeit in quite different ways—as the afterlives of what Gastón Gordillo has theorized as "rubble": the many layers of material, political, and affective debris that are left behind in the wake of capitalist dislocation.[2] Yet in addition to what Yarrow calls the "remains of the future," as important as this will be to making sense of both the presence *and* absence of lithium in Bolivia, I would add another framing, one that attends not so much to the afterlives of lithium extraction, but its *prelives*.

This is a way of conceptualizing what Palsson and Swanson would describe as the "geosocial" cycles of lithium through their multiple prefigurations.[3] As will be seen, these various prelives shape Bolivia's lithium project in a number of ways. On the one hand, tracing the prelives of lithium means recounting its iterations across time, space, and material transformation, each of which is embedded in a particular imaginary, a particular vision for how the wonders of chemical modernity would lead to a better future, a configuration that characterizes equally the invention and promotion of lithium-ion batteries. All of these prelives of lithium coalesce through each new application, building on each other and shaping how each is understood, justified, and sometimes abandoned.

But on the other hand, we must trace the prelives of lithium through a wider arc, one that encompasses not the history of the single mineral but the history of similar resource geopolitics, similar periods of time during which a resource shaped an epoch. In this case, it is impossible to understand the ethnography of Bolivia's ongoing lithium project without situating it in relation to centuries of extractivist entanglement. This is not merely, or most importantly, a question of broader context; even more, it is an ethnographic imperative. As I have argued elsewhere and as will be seen throughout the book, the fraught struggle over lithium industrialization in Bolivia is read in different ways by different actors and institutions *through* this longue durée of extractivism.[4] In this way, the remains of lithium's future in Bolivia are also filled with the debris from this longer resource chronology.

The next two sections of the chapter develop this dual temporality, first by recounting a selective history of lithium and then by curating an equally selective Bolivian resource genealogy. As will be seen, both of these resource narratives hold critical lessons for understanding the contested status of lithium in Bolivia today. The lithium destined for batteries and EVs today encodes a longer history of technological future-making and faith in scientific modernity. And like the lithium project, each earlier period of extractivism ramifies across multiple scales far beyond the borders of Bolivia itself, even if the environmental, social, and political costs are felt most directly by people and communities closest to the quotidian disruptures of exploitation, production, processing, and disposal.

The chapter then examines the emergence and evolution of interest in lithium resources at the Salar de Uyuni. As will be seen, this earlier history of lithium exploration and resource geopolitics in Bolivia coincides with similar processes taking place across the region, particularly in Chile. After tracing the prelives of lithium up to the beginning of what might be understood as ethnographic time, at which point my own research on different aspects of

Bolivia's lithium project provides much of the basis for analysis and reflection, the chapter then concludes with a short section that considers what it would mean to think of energy futures more generally as much through resource histories as through the possibilities for energy transition in the present.

A BRIEF ANTHROPOLOGICAL HISTORY OF LITHIUM

> Since sugar seems to satisfy a particular desire (it also seems, in so doing, to awaken that desire yet anew), one needs to understand just what makes demand work: how and why it increases under what conditions. One cannot simply assume that everyone has an infinite desire for sweetness . . .
>
> Sidney Mintz, *Sweetness and Power* (1985)

I had originally planned to anchor the ethnography of lithium industrialization in Bolivia—and its relation to wider processes of climate change mitigation, energy policymaking, and new resource geopolitics—in a more ambitious historical framework, one that would unpack each of the different iterations of lithium with the kind of cultural, political economic, and critical nuance that other anthropologists have brought to a wide range of resources and commodities, including palm oil, water, broccoli, tobacco, the matsutake mushroom, and, most iconically, sugar.[5] This ambition was also fueled by the knowledge that I would have access to one of the oldest and deepest bibliographic reservoirs in the world.[6] Furthermore, the literature lacked a wide-ranging and book-length social and political account of lithium. While the treatments by Bednarski and Kunasz, to which I will return below and in subsequent chapters, are useful, they are also limited in various ways;[7] a global history of lithium has yet to be written.

At the same time, as I moved through the different dimensions of lithium, it soon became equally clear that each demanded its own full-length study; in a sense, some of these *have* already been written. For example, the medical history of lithium has been well-recounted by the academic psychiatrist Walter Brown, among others, while Bednarski's book, although framed in broader terms, is actually a focused portrait of the contemporary global market for lithium-ion batteries, written by a lithium market analyst.[8] But even if the separate histories of lithium deserve to be brought together in a single comprehensive and critical volume, the current book is, alas, not the place for such an effort.

What follows, then, is a brief introduction to the material, medical, social, and military iterations of lithium, leading up through its use in batteries, although a significant part of the latter will be reserved for chapter 4, which examines Bolivia's attempts to develop the capacity to produce lithium-ion batteries, including the highly complex battery cathodes. Nevertheless, it is important to establish the outlines of this trajectory of lithium in order to better understand how Bolivia's lithium project takes place within a resource history that has unfolded in ways both linear and nonlinear. I describe this brief history as an *anthropological* one in order to emphasize both what seems to me to be most compelling about lithium as a cultural, social, and material artefact and to signal the aspects of this history, these prelives, that bear most directly on the ethnography of lithium at the heart of the book's narrative arc.

To return to the embarrassingly simple starting point—what is lithium? The story begins in Sweden with a twenty-five-year-old son of a wealthy bourgeois family named Johan August Arfwedson. Although he had completed a law degree at the University of Uppsala, his real passion was rocks and minerals, an interest he developed during his childhood wandering across his family's large estate in Skaraborg County.[9] Despite probable unease from his family, who expected Arfwedson to use his training in law to manage the family's many businesses and properties, he turned toward geology and chemistry, finishing courses in mining at Uppsala and then at the Royal College of Mines in Stockholm.

In 1817, while still a student at the Royal College of Mines, Arfwedson managed to secure a junior researcher position in the laboratory of Jöns Jacob Berzelius, a professor at the Karolinska Institute and one of the founders of modern chemistry. Using one of Berzelius's inventions, a portable blowpipe lamp, the eager Arfwedson conducted research on various minerals. While fusing a sample of petalite from the small island of Utö in the Swedish archipelago with potassium carbonate, Arfwedson obtained a salt that had an excess weight that could not be immediately explained. After further experiments, both Arfwedson and Berzelius concluded that the petalite contained an unknown alkali metal, which they named "lithium." Berzelius, who was well-connected with most of the leading chemists in Europe at the time, announced Arfwedson's discovery in correspondence and in different scientific proceedings, including in an article in volume 21 (1817) of the German *Journal für Chemie und Physik*.

Flush with excitement over the discovery of lithium, Arfwedson left Sweden on a multiyear scientific grand tour accompanied by his mentor

Berzelius, during which they paid visits to eminent researchers in England, France, and what was then Prussia. As recounted by Berzelius in his published travel diaries, he and Arfwedson had a number of interesting encounters, including during a carriage ride through a rural part of the Loire Valley: "Their fellow passengers were good-natured, inquisitive peasants who thought the Swedish language was a kind of French patois" and assumed that the young discoverer of lithium was a prince, given that "he was wearing in the cabriolet the same suit he wore on the streets of Paris."[10]

Later, after touring the silk and velvet factories in Lyon and visiting colleagues in Geneva and Zurich, where they both were elected honorary members of the Helvetian Scientific Society, the much wealthier—if much younger—Arfwedson decided that he had had enough of riding with "good-natured, inquisitive peasants" and bought a private carriage in Dresden, which he and Berzelius used for the remainder of their travels. Yet after returning to Sweden in 1819, Arfwedson was required to take over the management of a vast portfolio of his family's businesses, although he did equip a laboratory at his estate. He tried to maintain a parallel scientific career in subsequent years, even publishing a paper on uranium in 1822. But his research on minerals suffered and was later criticized for sloppiness—for example, in an infamous incident, Arfwedson mistakenly reported that he had derived silica from an analysis of Brazilian chrysoberyl when in fact it was the compound beryllia, which had been discovered twenty-five years before by the French chemist Louis Nicolas Vauquelin.

Arfwedson died in 1841 at the relatively young age of 49, having been retired from chemical research for almost two decades. Nevertheless, during the last year of his life, the Swedish Academy of Sciences awarded him a medal for his discovery of lithium, even though its later notice of commemoration was somewhat reserved. As the Academy put it, "One may venture to say that, because he was obliged to devote his time to the management of a considerable fortune, . . . the science to which he devoted himself in his youth lost much."[11]

But when Berzelius announced Arfwedson's great discovery in 1817 as "a new mineral alkali and a new metal," what did this mean? Lithium, like sodium, potassium, and several other elements from group one of the periodic table, are alkali metals, which means they share a number of characteristics, including being shiny, very soft, and so highly reactive that they must be contained in liquid solutions or in salt compounds. Lithium also has an extremely low density, which gives it a number of peculiar properties—for example, it can float on water. At the same time, as will be seen below and in chapter 4, lithium—with its very *low* density—is considered an ideal element in batteries because its different properties contribute to relatively *high* energy densities.

As an alkali metal, lithium has had a complicated history of medical and alimentary applications, also beginning in the nineteenth century. Around mid-century, lithium was first associated with the treatment of gout based on the experiments of the English physician and medical researcher Alfred Barring Garrod, who hypothesized that the condition was caused by abnormal levels of uric acid in the bloodstream. In his 1859 *The Nature and Treatment of Gout and Rheumatic Gout*, Garrod recommended the use of lithium salts to reduce uric acid, including for the treatment of "brain gout," which was theorized to be the cause of a range of mental illnesses, especially what was described at the time as "manic depression" (and later bipolar disorder).[12]

Throughout the rest of the nineteenth century, lithium was principally used for its apparent psychiatric benefits, at least for as long as the "brain gout" theory of mental illness was accepted.[13] Major centers for the psychiatric use of lithium included the United States, Denmark, and France, where a preparation called "Dr. Gustin's Lithium" was widely used in the south of the country, which explained the fact, as the French physician Roger Reyss-Brion put it, "that you don't have a lot of manic-depressives in Marseille."[14]

But as the brain gout explanation for mental illness rapidly disappeared in the early twentieth century, a medical transition influenced in part by the rise of Freudian psychoanalysis, so too did the therapeutic use of lithium. However, the use of lithium for dietary purposes became more prevalent as medical research began to link the use of table salt (sodium chloride) to a range of conditions, including high blood pressure and heart disease. Especially in the United States, doctors started to recommend the use of lithium as a replacement for salt at the dinner table.

The problem, as Walter Brown explains, is that "with a reduction in salt (sodium) intake, the body retains more lithium, setting the stage for lithium poisoning. In addition, the blandness of low-salt diets prompted some folks to ingest huge amounts of the lithium salt substitute."[15] This is why during the first half of the twentieth century, again, particularly in the United States, where the practice was widespread, the use of lithium as a condiment led to hundreds of documented cases of deaths due to "lithium toxicity," which eventually gave rise to what Brown describes as a national "salt substitute panic."[16] In 1948, the US Food and Drug Administration (FDA) banned the use of lithium as a substitute for salt, but also for another much more curious dietary use—as a key ingredient in soft drinks.

In the late nineteenth century, the commercial soda industry emerged, which brought industrial production capacity to bear on the preexisting taste for tonics and carbonated drinks with supposed medicinal properties. The most famous of these was Coca-Cola (originally invented as a cure for morphine addiction), which contained, among other things, cocaine and extract of kola nut. As Tristan Donovan describes in his social history of soft drinks, there were hundreds of different brands and patented recipes in these early decades when soda became ubiquitous, including one invented in 1929, at the dawn of the Great Depression, by a former ad executive turned amateur chemist named Charles Leiper Grigg.[17]

Seeing the market saturated with caramel-colored colas and fizzy orange sodas, including one that he had developed while working for the Vess Soda Company in St. Louis, Missouri, Grigg experimented with recipes for something different. Grigg eventually settled on a concoction that he christened the "Bib-Label Lithiated Lemon-Lime Soda," which contained lithium citrate, a compound that—despite the collapse of the "brain gout" theory of mental illness—had continued to be used on smaller scales to treat manic depression. Indeed, Grigg promoted the mood-altering effects of this new lithiated soda as an ideal beverage to raise spirits during the social and economic crisis of the 1930s. In 1936, he changed the name of this new elixir to 7 Up. (Grigg died in 1940, eight years before his company was forced to remove lithium from 7 Up by the FDA ban.)

After the invention of what became 7 Up, the US soda market was flooded with copycat "lithiated" drinks, most of them also lemon or lemon-lime flavored. Examples of these sodas from the 1930s and 1940s include Ace High Lithiated Lemon, Dice-Lucky Lithiated Lemon, Frisky Lithiated Lemon, Yawning Man Lithiated ("Does Not Say Wake-Up"), Manhattan Lithiated Gassosa, S.O.S. Lithiated Lemon ("First For Thirst"), Tune-Up Lithiated Lemon, Up 'N Up Lithiated Lemon, and Lithiated Champagne Dry Hi-Ball. Moreover, despite the 1948 FDA ban on the use of lithium, the actual prohibition was slow to take hold across the vast landscape of soda makers and distributors. For example, Isaly's (inventor of the Klondike Bar), founded by a Swiss immigrant dairy farmer in Ohio in the nineteenth century, was still offering the "lithiated lemon" flavor of its Isaly's Sparkling Mountain Air drink as late as 1950.

Coincidently, at the same time the FDA was taking steps to ban the use of lithium as a table condiment or additive in sodas, an obscure Australian psychiatrist named John Cade was methodologically experimenting with lithium compounds as a treatment for mental illness—not based on the discredited "brain gout" theory but on scientific studies he had carried out, including on himself.[18] In 1949, he published a landmark study on the use of lithium, a paper that "launched psychiatry's pharmacological revolution—the use of drugs to treat the mentally ill."[19] Cade began his research with lithium soon after returning to Australia as a former POW at the infamous Changi prison camp in Singapore, where the brutal conditions imposed by the Japanese exacerbated incidents of mental illness among the prisoners. Cade—as an Australian solider and medical doctor—had been put in charge of Changi's psychiatric ward, where he lacked the medical supplies or established protocols to treat many of his patients, including those suffering from severe depression.

During the 1950s and 1960s, lithium treatment for what became known as bipolar disorder was refined and standardized, particularly in order to maintain levels of lithium in the blood that are therapeutic but not toxic, a problem that had plagued Cade and others in the years after his 1949 research paper.[20] In 1970, the FDA finally relented and approved the use of lithium to treat bipolar disorder in the United States. As Brown argues, "Lithium enables most people with this formerly devastating condition to lead normal lives . . . Cade's breakthrough has prevented millions of suicides and salvaged an untold number of lives."[21]

As the writer Jaime Lowe puts it, someone who herself suffers from bipolar disorder, "I don't believe in God, but I believe in lithium."[22] Lowe explains that she actually traveled to the Salar de Uyuni, "one of the grandest, most delusional places of all," in order to "make a symbolic pilgrimage to the wellspring of [her] sanity." This is how she describes the experience: "After a few days of trekking, I stopped at a camp and slept in a building made of salt bricks—a lithium igloo. I sat in the nearby hot springs, in water naturally laden with high concentrations of lithium, and watched the steam rise on the moonshine horizon."

While the medical use of lithium was being developed and professionalized, yet another application of the alkali metal was put on terrifying display. During the Second World War, the two atomic bombs dropped on Hiroshima and Nagasaki by the United States were based on the principle of nuclear fission, in which energy is released by splitting the nuclei of atoms.

However, during the late 1940s and early 1950s, some of the same scientists who had been part of the Manhattan Project continued working on the nuclear program in order to produce weapons with much greater destructive capacities. To accomplish this, they proposed a two-stage process through which nuclear fission triggers a second reaction based on nuclear fusion, which requires a thermonuclear fuel. These scientists, working mostly at the Los Alamos National Laboratory in New Mexico, discovered that an isotope called lithium-6 made an ideal component for this fuel.

In March 1954, the first nuclear device with lithium as an essential material was detonated by the United States at Bikini Atoll in the Marshall Islands. The explosion was much greater than predicted because during the process of nuclear fusion, another lithium isotope, lithium-7, which was believed by the Los Alamos scientists to be inert, reacted as well, creating a blast and mushroom cloud with one thousand times more energy than the atomic bombs used by the United States against Japan.

As Parsons and Zaballa describe it, in their chilling history of the development and testing of thermonuclear weapons (also known as hydrogen bombs or H-bombs), the atmosphere-cleaving sight of this runaway, lithium-fueled explosion shocked the observers, most of whom were stationed on boats. The light from the blast was so intense, "they could see the bones in their arms, silhouetted against the glare." The bomb's "mushroom cloud rose to a height of over 114,000 feet, which is far up into the stratosphere"; the "atmospheric reflection of the blast was seen as far away as Okinawa, 2,600 miles distant—like an explosion in Boston that could be seen in Bogota [*sic*]."[23]

More importantly, of course, the fallout from the first lithium-fueled weapon was catastrophic for the ecosystems of the islands in the area, making life impossible for thousands of Marshall Islanders, who suffered the horrendous effects of radiation exposure, especially those on the islands of Rongelap and Utirik. The radioactive fallout from the blast eventually spread around the world through ocean and air currents, with high levels of radioactive particles being detected in Europe, India, and, ironically, the US Southwest, where the bomb itself had been designed.

As will be seen below, it is the nuclear application of lithium, more than any other, which first connects this chronology to the salares of South America, including the Salar de Uyuni. But before turning in the next section to the history of resource extraction in Bolivia, the second temporality through which the prelives of lithium are given shape and form, something must obviously be said about the origins of what has become the most globally consequential use of lithium—as the foundation for what Bednarski describes as the "new energy revolution."[24]

Although three scientists—John Goodenough, M. Stanley Whittingham, and Akira Yoshino—were awarded the Nobel Prize in Chemistry in 2019 for the development of the lithium-ion battery, it was not "another instance of a few inventors conjuring up a great idea, then cashing in," as the engineer and journalist Charles J. Murray puts it.[25] Instead, by the time the first lithium-ion battery was finally put on the market by Sony in 1991,

> It was as if a giant unseen hand had scooped up a nineteenth-century innova-
> tion and dropped it into the twentieth. It was not a software product, nor a
> semiconductor material, and it did not obey Moore's Law—which is to say
> that its cost did not drop by half every eighteen months. Moreover, it did
> not come from the mind of a single postadolescent billionaire. There was no
> eureka moment nor a tale of overnight success. As an invention, it was a closer
> cousin to the internal combustion engine than to the digital computer.[26]

Moreover, it is important to draw a distinction between lithium-ion bat-teries and electric vehicles themselves, which have a much longer history. Indeed, the first EV was launched by Thomas Parker as early as 1884 and by 1900, around 38 percent of all vehicles sold in the US were EVs based on Parker's design.[27] In 1913, Thomas Edison and Henry Ford collaborated on a new EV model with a nickel-iron battery in an attempt to improve its power and range. But as Murray's history shows, the idea of using batteries to power cars and trucks never really took off over the subsequent decades, for reasons that have to do with industrial competition, the much greater energy densi-ties of combustion engines in relation to these early batteries (including lead-acid batteries), and the kinds of geopolitical pressures associated with the twin rise of fossil fuel–based capitalism and mass democracy over much of the twentieth century.[28]

Nevertheless, by a twist of historic fate, the unfolding Age of Oil eventu-ally gave birth to the lithium-ion battery and its use to power a new genera-tion of EVs. In the early 1970s, just before the Saudi-dominated Organization of Arab Petroleum Exporting Countries (OPEC) decided to punish coun-tries for supporting Israel during the Yom Kippur War by declaring an oil embargo, Exxon instituted a research program to explore alternative sources of energy. The company had done a market analysis that predicted global oil production would hit a peak in the year 2000, then decline. Murray describes the impact of this internal report: "Researchers [within Exxon] were encouraged to look for oil substitutes, pursuing any manner of energy that didn't involve petroleum."[29]

In 1972, one of these Exxon researchers, the future Nobel laureate M. Stanley Whittingham, announced the invention of a battery that—for the first time—used lithium ions in one of the three major components (the electrolyte). As Murray explains:

> Whittingham's battery was unlike anything that had preceded it. It worked by inserting ions into the atomic lattice of a host electrode material—a process called intercalation. The battery's performance was also unprecedented: It was both rechargeable and very high in energy output. Up to that time, the best rechargeable battery had been nickel cadmium, which put out a maximum of 1.3 volts. In contrast, Whittingham's new chemistry produced an astonishing 2.4 volts.[30]

But despite the scientific innovation, Exxon, perhaps not surprisingly, never embraced the promise of the lithium-ion battery as a replacement for fossil fuels. Here the narrative takes a number of detours, marked by delays, missed opportunities, and the ever-present pressures of carbon geopolitics. At the same time, the two other eventual Nobel laureates were conducting research that led to advances in the two other main battery components, the cathode (Goodenough) and the anode (Yoshino). By the late 1980s, consumer electronics were becoming an even bigger global market, and Sony was looking for a new and lighter rechargeable battery for what became the Handycam. Using a prototype that incorporated the contributions of all three of the 2019 Nobel laureates, Sony scientists built the first commercially viable lithium-ion battery:

> Led by battery engineer Yoshio Nishi, Sony's team worked with suppliers to develop binders, electrolytes, separators, and additives. They developed in-house processes for heat-treating the anode and for making cathode powder in large volumes. They deserve credit for creating a true commercial product.[31]

During the 1990s, the success of Sony's lithium-ion battery led to it being rapidly adopted as the battery of choice in a wide range of personal electronics, from laptop computers to digital recorders to portable phones. In 1998, the Japanese automaker Nissan introduced a model called the Altra EV, the first to use a lithium-ion battery. Throughout the 2000s, a global EV market emerged, which included the upstart Tesla, whose 2008 Roadster model used a battery bank that contained almost seven thousand distinct lithium-ion battery cells. Nevertheless, despite the transformation in the global status of lithium in the wake of both advancements in EV design and the growing prominence of international climate policymaking, it is important to keep in mind that as late as 1994, only 7 percent of lithium produced in the world was used for batteries.[32] By 2008, its use in batteries had become the dominant application, and the search for new sources of lithium had become a major new resource geopolitics.

A GENEALOGY OF EXTRACTIVIST ENTANGLEMENT

If these various prefigurations of lithium shape the current lithium project in Bolivia and its fraught relations with the wider global energy transition, so too does a second chronology, not of lithium itself but of extractivist entanglement, the ever-present legacy through which resource extraction grounded an evolving mode of production through centuries of political conflict and social change. In tracing the prelives of lithium through this second temporality, it is important to emphasize both what threads through this admittedly condensed narrative and how this narrative—like that of lithium—provides essential context for understanding the ethnographic accounts to follow.

As will be seen, the current push to develop a lithium industry in Bolivia is marked by a number of factors that characterize other extractivist histories beginning in the early colonial period (and arguably before). This historical correspondence offers an important corrective to energy imaginaries in the present that all-too-readily frame both the lithium extraction-to-EV nexus and green energy technologies like lithium-ion batteries more generally in millenarian terms. Put another way, the longer chronology of resource geopolitics within which lithium industrialization in Bolivia today must be located shows how resource extraction is always linked to—and justified by—visions of a different and better future, one made uniquely possible by the extraction of each new raw material or "critical" mineral.

In addition to the presence of different iterations of resource utopianism and the role of technological development as a mechanism of this transhistorical imaginary, a genealogy of extractivist entanglement reveals other interconnecting variables, including the use of various forms of violence (economic, political, military, symbolic); the politicization of scarcity as both an economic category and basis for geopolitical conflict; and the constant presence of resistance.

FIGURE 4. "Bolivia Will Not Be a Hacienda for the Gringos!" La Paz Teachers Union, 2009. Photo by author.

And although resistance takes different forms—technological, material, social, and so on—here it is the political expression of resistance that is most revealing about how the *longue durée* of extractivism informs the ethnographic analysis of Bolivia's lithium ambitions. Yet more often than not across this history, it is the *terms of entanglement* rather than extractivism itself as a mode of production that shapes mobilization and animates calls for change, including revolutionary change.

But before we can appreciate the key distinction between resistance to certain forms of entanglement and resistance to extractivism, we must begin this genealogy with what has become a dark icon in the annals of anti-extractivist critique: silver mining in Potosí, the capital of the administrative department in which Bolivia's lithium facilities are located. For the Uruguayan essayist and historian Eduardo Galeano, the silver mine of Cerro Rico ("Rich Mountain") was the ur-symbol for global plunder, extractivist social suffering, and forced resource dependency.[33] Indeed, the mine of Potosí and its place in global economic history are used by Galeano as a metaphor for capitalism itself, which depends on the extraction of wealth from violently opened veins until they inevitably, tragically, bleed out. Drawing on the archival research of the economic historian Earl Hamilton, Galeano argues that the sixteen million kilograms of silver extracted from Spain's colonial mines in Latin America between 1503 and 1660, mostly from Potosí's Cerro Rico, laid the foundation for the rise of both mercantile colonialism and early modern Europe itself.

The small town at the base of the silver-rich mountain, which had been incorporated into the Inca Empire before the arrival of the Spanish, grew in unprecedented ways as "captains and ascetics, knights and evangelists, [and] soldiers and monks came together in Potosí to help themselves to its silver."[34] These were in addition to the thousands of *corvée* laborers, who were forced to work in the mines when the Spanish colonial administration imposed a labor regime that came to be known as the *mita*.[35] By the mid-seventeenth century, Potosí had more inhabitants than Madrid, Rome, or Paris: a 1650 census "gave Potosí a population of 160,000. It was one of the world's biggest and richest cities, ten times bigger than Boston—at a time when New York had not even begun to call itself by that name."[36]

Potosí's mining, political, and ecclesiastical elite built a city at thirteen thousand feet with roads literally paved with bars of silver; and while the mineral wealth of Cerro Rico fueled centuries of Spanish and wider European colonialism and provided the financing for what would become global capitalism, these entanglements also brought this wider global system of extraction to the "Imperial City" itself: "silks and fabrics came from Granada, Flanders, and Calabria; hats from Paris and London; diamonds from Ceylon; precious stones from India; pearls from Panama; stockings from Naples; crystal from Venice; carpets from Persia; perfumes from Arabia; [and] porcelain from China."[37]

Yet well before Bolivian independence in 1825, Potosí had been abandoned by the Spanish colonial regime, which had long since transferred its extractivist operations to locations more favorable to global trade, especially to Mexico, where the silver mines were much closer to the Caribbean ports. Thus began what Galeano describes as Potosí's slow "descent into the vacuum," a long process of colonial *disentanglement* that "left Potosí with only a vague memory of its splendors, of the ruins of its churches and palaces, and of 8 million Indian corpses"[38]—people who perished over the centuries in a dizzying array of mining accidents or from the ravages of silicosis, which even today strikes down many Bolivian miners still in their thirties and forties. Writing in the early 1970s, Galeano quotes an elderly woman he met in Potosí, whose vision of the crumbling city echoes the skepticism with which many people and social movements in Potosí more recently have regarded the state lithium project: her hometown was "the city which has given most to the world and has the least."[39]

If the shadow of Cerro Rico and the ghosts of the millions who were "eaten" by the mine continue to haunt Bolivia's unfolding lithium project, so too does a resource conflict that took place toward the end of the nineteenth century.[40] At the time of Bolivian independence in 1825, its territory included roughly six hundred kilometers (km) along the Pacific Ocean from the Peruvian border to the north to the Chilean border to the south. This region was originally a province of the Potosí Department, which became the Littoral Department (Coast Department) in 1867 during the government of Mariano Melgarejo.[41] The new department was further divided into two provinces: La Mar, whose capital was Cobija, located on the Pacific coast and Atacama, whose capital was San Pedro de Atacama in the Atacama Desert to the east. As Mesa, Gisbert, and Mesa Gisbert explain, one of the peculiarities of the Littoral Department, which had a population of about fifteen thousand people, was that a vast majority considered themselves Chilean rather than Bolivian.

The reason such an "inhospitable region" was so important, both for Bolivia and the rest of the world, was that it contained two resources of global importance during the nineteenth century: guano and saltpeter.[42] Although bird colonies are obviously found in many locations, the bird colonies along the coastline where the Atacama Desert meets the Pacific Ocean are marked by an extraordinary feature: despite the presence of seawater, the offshore islands and rocky coastal outcrops that host thousands of birds— including the Peruvian pelican, Peruvian booby, and guanay (a kind of cormorant)—are places where bird feces accumulates over many years and where there is very little, if any, precipitation. This climatological oddity allowed vast deposits of guano to build over time. As early as the Inca period, but likely much earlier, Indigenous farmers had used guano to fertilize their fields because it contains high levels of different elements that enhance plant growth, especially nitrogen.[43] Although many of the richest guano deposits were controlled by Peru, especially the Chincha Islands, Bolivia's Littoral Department also had a number of important guano fields.

At the same time, the interior of Bolivia's Atacama Desert provided another source for nitrate compounds: a sedimentary rock called *caliche*, which contains sodium nitrate, also known as saltpeter. Although both guano and saltpeter were highly valued during the nineteenth century in the

United States and Europe as agriculture became more industrialized, both were also used for another application of even greater geopolitical consequence: as gunpowder for dynamite and military armaments. By the nineteenth century, the use of saltpeter for gunpowder had been well established for centuries. Gunpower, in turn, was the basis for imperial expansion and resource conflicts, a history that made the search for and control of saltpeter, like its nitrogenous cousin guano, essential for European and later US colonialism. As Cushman explains:

> The saltpeter trade . . . inspired constant political intrigues and an occasional nitrogen war. In 1757, during the earth's first true world war, the British East India Company orchestrated the overthrow of the nawab of Bengal and installed a native ruler who gave the company a near-monopoly over South Asian saltpeter exports. This action not only opened the way to British colonialism in India and gunboat diplomacy elsewhere, it also enabled the British to starve its rivals of [saltpeter] and tighten its dominance over the trans-Atlantic slave trade through sales of guns and gunpowder.[44]

The over-extraction of guano throughout much of the nineteenth century in Peru and Bolivia, coupled with a catastrophic earthquake in 1868, a magnitude 9.0 "megathrust . . . [that] caused a tsunami comparable to the 2004 Indian Ocean event" and destroyed much of the coastal guano industry, heightened the importance of Bolivia's inland saltpeter reserves.[45] Although nominally regulated and taxed by Bolivia, much of this industry had been developed and capitalized by foreign companies, the most important of which was the Antofagasta Nitrate & Railway Company, owned by a consortium of British and Chilean shareholders. In order to cover major losses to its coastal region, Bolivian politicians from the Littoral Department managed to convince the central government to levy a new tax on exports, a tax that the Antofagasta Company refused to pay.

In 1879, under the imperial gaze of both London financiers and the British government, which had maintained close historical and economic ties with Chile since before its independence, the Chilean military launched a preemptive invasion of both Bolivia and Peru, with whom Bolivia had signed a mutual defense treaty. The resulting "War of the Pacific (1879–84) was one of the largest armed conflicts ever fought in the Americas and provided a preview of the massive wars fought over phosphate, petroleum, *Lebensraum*, and other resources during the twentieth century. It violently confirmed the significance of nitrogen compounds to global history during the nineteenth century."[46]

In the aftermath of what ended up being a complete Chilean military victory, Bolivia lost its entire Pacific coastline, which was occupied and then absorbed into Chilean national territory through the humiliating 1884 Treaty of Valparaiso, an initial peace agreement between Bolivia and Chile that was made permanent twenty years later through the euphemistically described Treaty of Peace and Friendship between the two countries. In addition to ceding its Pacific coastline to Chile, a historical tragedy for Bolivia that continued to shape national politics throughout the twentieth century and into the early twenty-first via various forms of the "Sea for Bolivia" campaign, Bolivia also lost its saltpeter mines when Atacama became part of Chile.[47] Over a century later, in the early years after the Augusto Pinochet dictatorship (1973–1990) ended, Sociedad Química y Minera de Chile (SQM), led by Pinochet's son-in-law Juan Ponce Lerou, developed the first industrial lithium industry based on brine evaporation, with the most important facilities located at the Salar de Atacama.

The national trauma of the War of the Pacific, a resource conflict precisely over the terms of extractivist entanglement, continues to shape Bolivia's lithium project in a number of ways. On the one hand, the circumstances of the war and the still-open wound of its demeaning resolution are experienced by Bolivians across the otherwise violently divided political spectrum as a legacy in which resource geopolitics are fundamentally associated with national disgrace and loss. This means that even if the nascent lithium industry in Bolivia is directed by the state, any proposal to partner with foreign companies—to build facilities, transfer technology, or provide project management—is viewed through this dark legacy.

But on the other hand, and more practically, the enduring resentment over the War of the Pacific in Bolivia, a resentment that has been stoked over the decades by a wide range of governments, from the dictatorships of the 1960s and 1970s to the MAS administrations of more recent years, explains why Bolivian scientists, engineers, and energy policymakers would never consider asking for assistance from their much more experienced Chilean colleagues, at least not officially. Even if Bolivia will never manage to convince Chile to return its lost coastal territories, it might very well manage to develop an industrial lithium capacity that equals or surpasses that of Chile.

A final conflict over the terms of extractivist entanglement, one that also continues to shape the contours of Bolivia's lithium ambitions, convulsed the country in the waning years of the neoliberal ancien régime (1982–2005). This was the period after the restoration of democracy during which the

Bolivian government infamously imposed a series of radical free market reforms, an extended period marked by pervasive "clientelism, corruption, and exclusion."[48]

In the midst of imposing this economic "shock doctrine," Bolivia's neoliberal governments undertook a series of social reforms through which rights frameworks, especially Indigenous rights, were given much more legal and political prominence. This took place at the same time that the country doubled down on its commitment to the privatization of major industries, including oil, gas, and mining. This effort to combine human and Indigenous rights reforms in Bolivia with market capitalism and its associated dislocations was not a historical aberration. As Jessica Whyte has convincingly argued, the "morals of the market" have always demanded that accumulation by dispossession be justified in the language of rights.[49]

But these arguments within Bolivia were no match for other visions of social and economic justice. These included those anchored in the country's long tradition of Trotskyist labor politics and more recent lines of political discourse that have sought to revendicate Bolivia's historically marginalized Indigenous populations by privileging alternative social structures, economic networks, and temporalities. It was precisely at the moment in which the long-standing tension between the country's neoliberal political economy and the revolutionary ferment of its diverse social and political movements could no longer be sustained that the system fractured.

In 1997, Hugo Banzer Suárez—the country's military dictator from the 1970s who we will meet again below—was elected to the presidency, with business administrator and former minister of finance Jorge Quiroga Ramírez as his vice president. However, suffering from terminal lung cancer, Banzer resigned from office in 2001 and Quiroga assumed the presidency. Despite the fact that the next presidential elections were scheduled for 2002, Quiroga pressed negotiations with Chile over a proposal to build a gas pipeline to Bolivia's former coastal territory, where privately owned Bolivian gas would be processed and then shipped abroad, primarily to markets in Mexico and the United States. Before the controversial deal could be finalized, new elections saw another return to power, this time of the arch-neoliberal Gonzalo Sánchez de Lozada, known as Goni, who had served as president between 1993 and 1997.

Sánchez de Lozada's government was racked by social unrest from the beginning—over land reforms, tax increases, and relations with labor unions, including the surging coca growers' union headed by Evo Morales. In the midst of these disputes, Sánchez de Lozada returned to the proposal to build a gas pipeline through Chile to the Pacific Ocean, a proposal that now included offers by the Chilean government to both oversee the refining of privatized Bolivian gas on Chilean territory and to buy a portion of the refined gas itself. When the gas pipeline deal became public knowledge, it was the final spark that ignited what became known as the Gas War, a turning point that also "pushed Sánchez de Lozada into the abyss."[50]

In late 2003, widespread social mobilizations against the proposed gas deal, which included a general strike called by the Central Obrera Boliviana (COB), Bolivia's national trade union federation, led the national government to turn the military against its own citizens after having declared martial law. The most serious confrontations took place in the "rebel city" of El Alto on the sprawling altiplano above La Paz,[51] where dozens of protestors were killed by Bolivian soldiers and police in a dark moment that soon came to be known as Black October. Key members of the government withdrew their support for Sánchez de Lozada and the government quickly collapsed. The proposed gas pipeline deal with Chile also obviously collapsed. It would, however, play an important role in the MAS campaign of 2005, in which Morales promised to fully renationalize the Bolivian oil and gas industry within months of taking office. Not only did the renationalization take place (in May 2006), but it was defended as the beginning of a broader transformation of the country's resource politics, a transformation that would later shape the lithium project.

More concretely, the Gas War of 2003 was marked by the same factors that underlie the other signal conflicts over the terms of extractivist entanglement in Bolivia. Leaving aside the ever-provocative—even socially intolerable—place of Chile at the heart of the crisis, two other aspects must be emphasized. First, the technologically specific question of refinement took on wider political and symbolic significance. It was not just that natural gas would be exported from Bolivia by private companies; rather, what infuriated many was the fact that the refinement process would both add significantly greater value to a Bolivian natural resource and take place beyond the control of the Bolivian state.

And second, the Gas War, like the other resource conflicts, reveals yet again an enduring truth, one with consequences in particular for the broader environmentalist critique of the energy transition: that it is not usually resource extraction itself that is of primary concern, except for the important exception of some communities (but not all) closest to the sites of exploitation and production. Instead, the thread that runs through the entire resource chronology, one that leads directly to Bolivian's ongoing lithium project, is the historical and conceptual significance of entanglement. Who defines the terms of entanglement? Who benefits from the entangled webs of resource geopolitics? And which kinds of entanglement trigger resistance? Is it possible that lithium production in Bolivia will one day come to resemble precisely those categories of entanglement that spark the fiercest resistance, entanglements in which the open veins of Bolivia bleed, not for Bolivians themselves, but to nourish economies, visions, and empires far from Bolivia's borders?

PROJECT EROS AND THE 1976 "LITHIUM SAFARI"

Having traced the prelives of lithium—selectively—through both their material and geopolitical prefigurations, let me now turn to a more conventional historical narrative, one that focuses on the emergence and evolution of interest in lithium resources at the Salar de Uyuni. This narrative returns us to the 1970s, a time of dictatorship, Cold War neocolonialism, nuclear proliferation (despite the 1970 Non-Proliferation Treaty), and crisis within the global capitalist system.

In 1971, Hugo Banzer Suárez—then *Colonel* Banzer—ousted another military leader in a bloody coup d'état that continued a period of alternating military rule in Bolivia that had begun in 1964. Despite what one would imagine given the wider history of military governance in Latin America, not all leaders in Bolivia during the period 1964–1982 were equally pro-US and therefore equally committed to a politics of procapitalist anticommunism. Banzer, however, whose reign came to be known as the *Banzerato*, the dreaded Time of Banzer, was not an exception to the general pattern: quite the contrary. During the Banzer dictatorship, his government suppressed or banned labor unions; used torture and disappearances against activists; actively participated in the CIA-funded Operation Condor, the coordinated program in South America to crush leftwing politics; and ensured the country's natural resources were available for foreign appropriation, especially by US companies.

It was at this moment, the early years of the Banzer dictatorship, when Bolivia's lithium resources become enmeshed in this web of authoritarian rule, US interventionism, and the use of advanced technology to gain eco-

UNITED STATES
DEPARTMENT of the INTERIOR
★ news release

OFFICE OF THE SECRETARY Forrester 343-4646

For release: SEPTEMBER 21, 1966

EARTH'S RESOURCES TO BE STUDIED FROM SPACE

Project EROS was announced today by Secretary of the Interior Stewart L. Udall. EROS (Earth Resources Observation Satellites) is a program aimed at gathering facts about the natural resources of the earth from earth-orbiting satellites carrying sophisticated remote sensing observation instruments.

"Project EROS", said Udall, "is based upon a series of feasibility experiments carried out by the U. S. Geological Survey with NASA, universities, and other institutions over the past two years. It is because of the vision and support of NASA that we are able to plan project EROS."

Udall said that "this project will provide data useful to civilian agencies of the Government such as the Department of Agriculture who are concerned with many facets of our natural resources. The support of these agencies is vital to the success of the program."

The Interior Secretary said that "the time is now right and urgent to apply space technology towards the solution of many pressing natural resources problems being compounded by population and industrial growth."

Udall said that "the Interior Department program will provide us with an opportunity to collect valuable resource data and use it to improve the quality of our environment."

FIGURE 5. First page of the press release announcing the launch of the EROS program, 1966.

nomic advantage. Five years before Banzer took power, the US announced the launch of a new program, one that went largely unremarked at the time—and arguably since—amid both the national focus on the lunar space program and the growing US military involvement in Vietnam. In September 1966, the US Department of the Interior began something called "Project EROS," which was described as a way to harness space technology to allow the US to map the Earth's natural resources so that the US could better respond to "problems being compounded by population and industrial growth."[52]

The acronym EROS, after the Greek god of love (Aphrodite's son), stood for "Earth Resources Observation Satellites." The objective of the program, which was designed to launch a series of satellites into orbit, was clear: it would give the US a competitive advantage over its political economic rivals in the world by providing a technologically proprietary dataset about the Earth's natural resources, data that would also form the basis for the kinds of "alliance-building" and interventions that had marked US foreign policy throughout the Cold War. As the Project EROS press release put it, the satellite surveillance program "would provide technological support for the continuation of our society of 'plenty' for generations to come. EROS will be just the beginning of a great decade in land and resource analysis for a burgeoning population."[53]

Despite this initial enthusiasm, it took six more years before the first surveillance satellite was launched, by which time Project EROS had come under the joint control of the US Geological Survey (USGS) and the US National Aeronautics and Space Administration (NASA). One of the leaders of the EROS program was William D. Carter, a geologist who would go on to become one of the most vigorous advocates within the US government for the use of space technology to search for minerals of strategic political and economic importance.[54] He had also worked for years in Chile conducting field surveys of potential mineral deposits.

In July 1972, the first surveillance satellite, named ERTS-1, built by private US defense contractors, was put into operation. Between July 30 and October 30, 1972, ERTS-1 overflew the central Andes and obtained images of the Bolivian altiplano, including the Salar de Uyuni. These images revealed signatures of potential mineral resources in addition to hydrological data about critical drainage basins and detailed information about volcanic geomorphology.[55] After EROS identified locations that might be useful for reinforcing a "society of plenty" in the US, the procedure was then for the US government to sign agreements between the USGS, NASA, and foreign counterparts, agreements that would give the USGS privileged access to sites of interest, where more detailed field research could be conducted to confirm the actual extent and composition of the identified deposits.

Given the potential importance of the Salar de Uyuni (and also of the Salar de Coipasa, which showed similar signatures in the ERTS imaging), the US government quickly financed a program in Bolivia (in August 1972) through a partnership between the USGS and Bolivia's national geology service. The director of what was called the ERTS-Bolivia Program was Carl Edward Brockmann Hinojosa, a senior state geologist from Cochabamba who had worked for most of his career in the oil industry. It was Brockmann—who gave a paper in 1974 at the NASA Goddard Space Flight Center on what the ERTS satellite had revealed about potential Bolivian resources—who coordinated the first field surveys for minerals at the Salar de Uyuni, the most important of which took place in April and September of 1976.

Although William D. Carter was one of the leaders of Project EROS—whose data analysis center was (and still is) located in a remote field north of Sioux Falls, South Dakota—he was also responsible for conducting follow-up field research in his own right. This was done in order both to refine the techniques through which initial information from the surveillance satellite would be used to confirm the nature and extent of resources and to establish (or reinforce, as with Bolivia) a US governmental toehold in countries of strategic political economic interest. Eager to advance the US mission of what he described as satellite "mineral finding," Carter arrived in Bolivia in April 1976 and traveled to the Salar de Uyuni with a young Bolivian state geologist from La Paz named Raúl Ballón Ayllón.[56]

Carter and Ballón collected two samples of brine from the southeastern Salar de Uyuni, which were then sent to the USGS's research chemistry department in Reston, Virginia. In June of 1976, Shirley L. Rettig, a USGS chemist, issued a report on the two brine samples. The first returned a lithium concentration of 490 parts per million (ppm), which was significant—but the second had an extraordinarily high concentration of 1,510 ppm. As a later report noted, the lithium concentrations in the sample brine from Bolivia were orders of magnitude higher than the strategically important brines of Nevada and California, whose value was still primarily associated with the use of lithium in nuclear weapons.[57] For example, a brine deposit in Clayton Valley, Nevada, had concentrations of about 300 ppm of lithium, while another, at Searles Lake, California, only averaged 70 ppm.

The discovery of the "existence of anomalously high"[58] levels of lithium at the Salar de Uyuni led to a flurry of activity by the US government soon after receiving the June 1976 report. In September 1976, a second expedition was organized to survey the lithium reserves of the Salar de Uyuni, only this time the team was bigger and more diverse, something that reflected its importance to the US and, by extension, Bolivia, at a time in which the Banzer dictatorship was at the height of its eagerness to placate both US political and economic interests in the waning months of the Republican Ford administration.

In close coordination with John H. Curry, Regional Minerals Attaché at the US Embassy in La Paz, the September 1976 lithium survey consisted of the following members: George E. Ericksen and James D. Vine, two senior USGS geologists who were leading experts in South American salar geology, including in brine formation and chemistry; Ballón, the Bolivian state geologist; and, critically, two highly experienced geologists from the largest private US lithium companies, Gerald Blanton from the Lithium Corporation of America (Lithco) and Ihor Kunasz from the Foote Mineral Company.[59]

Over the course of the entire month of September, the five geologists conducted a comprehensive survey of the Salar de Uyuni, with smaller surveys done at both the Salar de Coipasa and Salar de Empexa. Using gasoline-powered saws with a "specially designed drill stem equipped with tungsten carbide teeth,"[60] they cut thirty-nine fifty-centimeter bore holes across the entire Salar de Uyuni along bisecting north-south and northwest-southeast axes. At twenty of the bore holes on the salar, the geologists took particularly large (two thousand milliliters) brine samples. This was done so that the private geologists—Blanton and Kunasz—would have their own brine to analyze once they returned to their corporate laboratories in the US.

The results of this major survey—which the Foote geologist Kunasz would later describe in a memoir as the "lithium safari"[61]—were spectacular: not only did the analysis confirm the "anomalously high" concentrations of lithium in the Bolivian brines, it also revealed that the lithium reserves were likely to be the largest in the world, given that the Salar de Uyuni alone covered an area of 10,500 square kilometers. However, despite the fact that the geologists had uncovered the most significant lithium deposit on the planet, all the main protagonists would be left bitterly disappointed. For reasons described in the next section, this discovery *did not* directly lead to the next steps likely imagined by the Regional Minerals Attaché in La Paz, the USGS, and the satellite miners of Project EROS, not to mention the private geologists for whom the entire resource mapping project was ultimately directed: namely, the beginning of intensive lithium extraction in Bolivia under the supervision of private US lithium companies.

And what about Project Eros and the grand US strategy to use satellite technology as a much earlier and quite different tool for what Shoshana Zuboff would later describe as "surveillance capitalism"?[62] In 1975, the ERTS satellite was renamed Landsat 1, at the same time that EROS was retained as the acronym for the USGS data center in South Dakota. The resource mapping project was expanded and continued through a series of Landsat satellites; the most recent, Landsat 7, was launched in 2021. As NASA explains, in euphemistic language that obscures the origin and ongoing rationale for the resource mapping project, "Landsat satellites have continuously acquired images of the Earth's land surface and provided an uninterrupted data archive to assist land managers, planners, and policymakers in making more informed decisions about natural resources and the environment."[63]

Perhaps most important for the contemporary study of lithium extraction and the energy transition, the Landsat satellites also play a critical role in tracking key indicators of climate change, including shifting weather patterns, glacial melt, and national "adaptations" to these changes. In this way, the technology that was first used to reveal Bolivia's lithium reserves is now deployed as part of the wider strategy of climate change mitigation. But both the political economic and technological continuities at the heart of the Landsat program suggest that climate change mitigation and the violent struggle to ensure a "society of 'plenty' for generations to come"—but only for certain societies—are tightly and inescapably intertwined.

FROM "COLONIAL SCIENTIFIC RESEARCH" TO THE STIRRINGS OF RESOURCE REGIONALISM

Even if the early US political and scientific presence in Bolivia formed the basis for later stages in Bolivia's lithium project, especially its focus on the southern and southeastern zones of the Salar de Uyuni, where the highest lithium concentrations were discovered, and, more broadly, continued a long-standing policy of economic interventionism (something made brutally clear during the 1973 CIA-backed ouster of Chile's Salvador Allende), it was not the only state actor to take an interest in Bolivia's mineral resources during these years.

In 1974, two years after the ERTS-Bolivia Program had been launched, the French government signed an agreement with the Universidad Mayor de San Andrés (UMSA) in La Paz to collaborate in the study of Bolivia's mineral deposits, especially those at the Salar de Uyuni.[64] The French agency that was charged with developing this collaboration was an institution whose origins—like those of Project EROS—provide clear and compelling evidence of the objectives behind France's sudden interest in Bolivia. The Office de la

Recherche Scientifique et Technique Outre-Mer (ORSTOM) was created by the wartime French Vichy regime in 1943 under a different name: the Office for Colonial Scientific Research.[65] The purpose of ORSTOM was to promote scientific research as a pillar of French colonial expansion, which included the search for resources of strategic importance to France.

The point person for ORSTOM's foray into the lithium stakes in Bolivia was a young Alsatian geochemist named François Risacher, whose research would come to play a critical role in expanding the knowledge about lithium in Bolivia, including establishing 8.9 million tons as the likely extent of lithium reserves, a figure that was accepted from 1989 until the landmark survey project undertaken by Bolivian researchers between 2013 and 2018.[66]

Risacher, who was only in his mid-twenties when he arrived in Bolivia, had recently graduated from the École Nationale Supérieure de Géologie. As an ORSTOM researcher, he would specialize over the course of his long career in the study of the salares of Bolivia and Chile. Among other important findings, Risacher and his research collaborators from UMSA and the Bolivian state geology service, including Oscar Ballivián, Hugo Alarcón, and Ricardo Morales, were the first to establish the importance of the Río Grande drainage basin as the principal source of the brine with the highest lithium concentrations.

While both the French and US projects worked with their Bolivian counterparts to explore the resources of southwest Bolivia, sometimes even in collaboration with each other, the Banzer regime was doing its part to get Bolivian institutions ready for what it hoped would an important new source of international investment.[67] As early as 1974, having already authorized the establishment of both ERTS-Bolivia and the more recent ORSTOM agreement, Banzer signed a "supreme decree" declaring the region of the salares a national "fiscal reserve." Despite its chilling name, a supreme decree in Bolivia is basically an executive order, a legal category that allows the government (of whatever kind) to take action while the congress debates ratifying the decision through legislation. But in designating the salares—most importantly the Salar de Uyuni—a fiscal reserve, the Banzer government was both invoking a traditional state mechanism for asserting control over the country's resources and laying the groundwork for jurisdictional conflicts over the lithium project that still have not been resolved.[68]

In 1976, the Banzer government (also through a supreme decree) went a step further and created a program of infrastructural development for the salar region under the supervision of a state company (QUIMBABOL) controlled by the Bolivian armed forces. However, before this program could be put into place, the entire nascent lithium project in all of its different facets came to a halt. Between 1978 and the early 1980s, Bolivia entered a period of dictatorship and oppression that was even darker than the years that preceded it. A series of coups d'état (including the notorious Cocaine Coup of 1980, which brought Luis García Meza to power) finally gave way to the restoration of democracy in Bolivia after almost twenty years.

For our purposes, the next stages in the chronology can be taken together, since they all unfolded under different neoliberal governments, a series of developments that culminated in the rise of Evo Morales and the MAS and the beginning of a new lithium temporality in Bolivia. In 1985, the government of Hernán Siles Zuazo returned to the question of lithium. It created a state "industrial complex," the Complejo Industrial de los Recursos Evaporíticos del Salar de Uyuni (CIRESU), whose specific mandate was to act as the agent of the Bolivian government for contracts with foreign companies and investors who wanted to develop the "evaporitic resources of the Salar de Uyuni."[69] This model for lithium industrialization, in which the Bolivian state grants extraction rights to foreign companies in exchange for taxes and royalties, was soon put to the test in a drawn-out episode with lasting consequences.

In 1987, the mining minister in the Siles Zuazo administration, Jaime Villalobos, bypassed the regulatory process and made a direct no-bid offer to Lithco (now known as FMC-Lithco), whose company geologist (Gerald Blanton) had been a member of the 1976 "lithium safari," to begin extraction operations at the Salar de Uyuni.[70] In 1989, after two years of negotiations between FMC-Lithco and the Bolivian government, the terms of a draft contract were announced by the incoming government of Jaime Paz Zamora; they gave FMC-Lithco extraction rights over the entire salar fiscal reserve for a period of forty years.

But in a pattern that would be repeated throughout the subsequent history of lithium in Bolivia, the no-bid deal with FMC-Lithco triggered widespread resistance from a number of regional institutions, including the Comité Cívico Potosinista (COMCIPO), a "civic" social movement located in the former "Imperial City" of Potosí. Even more important, the FMC-Lithco proposal was opposed by a regional peasant movement with much more influence in the salar region than the civic organizations located in the departmental capital: the Federación Regional Única de Trabajadores Campesinos del Altiplano Sur, or FRUTCAS.

In a rapid-fire series of developments, in late April 1990 the contract with FMC-Lithco was finally signed by the Paz Zamora government, only to have the same government annul the contract less than a week later amid massive street mobilizations and even hunger strikes, both in Potosí and near the Salar de Uyuni. To placate FMC-Lithco, however, the Bolivian government announced that the process would continue through a public tender and that FMC-Lithco would be in a strong position given its existing commitment to the project. In early 1992, after two years of public debates and scientific forums, the consideration of different proposals, and continuing political mobilizations around lithium extraction, the Paz Zamora government announced—not surprisingly—that it had chosen FMC-Lithco to receive the contract among the three finalists, which included SOQUIMICH, a subsidiary of the private Chilean company SQM, which would emerge as one of the world's largest producers of lithium during the 1990s, and COPLA, a Bolivian mining company founded in 1988 that specialized in the production of boron compounds.[71]

But just days after the new contract with FMC-Lithco was signed, the process was upended yet again. The Bolivian congress intervened, at the same time that FRUTCAS also continued to pressure Paz Zamora. Among other things, FRUTCAS objected to the fact that only 2 percent of projected earnings by FMC-Lithco would be returned to the salar region through royalties.[72] And then the Bolivian congress, in a stunning move, raised the value added tax from 10 to 13 percent and applied this retroactively to the FMC-Lithco contract, despite language that was supposed to protect the US company from any subsequent changes to Bolivian law during the contract's forty-year duration.

Over the next year, as new elections ushered in the government of Gonzalo Sánchez de Lozada, the position of FMC-Lithco remained uncertain, apart from the fact that the company was obviously unhappy with the sudden tax increase. Unbeknownst to most of the different actors on the Bolivian side, however, FMC-Lithco had been hedging its bets on lithium development in the region since early 1991. While debates raged within Bolivia over the fallout from the initial no-bid contract, FMC-Lithco had quietly signed another deal with the province of Catamarca in Argentina, which gave the company the exclusive rights to explore and industrialize the resources at the Salar del Hombre Muerto, another salar with immense lithium reserves.[73]

In November 1993, FMC-Lithco announced that it was abrogating the deal with Bolivia based on breach of contract. More likely, the company had begun working behind the scenes to prepare for this shift almost immediately after the Bolivian congress announced the tax increase, which they took as a worrying sign of things to come at a moment in which lithium extraction from brine at industrial scales was still very much a risky proposition. FMC-Lithco simply moved the operations and financing planned for the Salar de Uyuni to Argentina, where the company managed to begin producing battery-grade lithium carbonate for the rapidly expanding lithium-ion battery market in only three years.[74]

The consequences of the collapse of the FMC-Lithco deal were enormous. On the one hand, this history would shape global discourse around Bolivia's lithium ambitions through multiple periods and economic models, including those at the heart of MAS's later strategy of "productive" sovereignty. Even though the value of Bolivia's lithium would over the years continue to attract the increasingly fevered interest of a range of state and private companies, the legacy of 1992–1993 would remain as a warning, something to haunt the dreams of risk managers and market analysts.

But the lessons of 1992–1993 were quite different for the social movements whose resistance to the deal had played an important role in pushing it off the rails. As will be seen in later chapters, both COMCIPO and FRUTCAS had put Bolivia's national government on notice: *any* proposal to extract and industrialize lithium at the Salar de Uyuni would have to meet a number of political and economic criteria, including state control and the promise to redistribute earnings between the city of Potosí and the local communities around the salar, a demand that would eventually create friction *between* the different social movements. Nevertheless, this signal moment of resource *regionalism*, which must be distinguished from resource nationalism, would become a guide to action for both COMCIPO and FRUTCAS, one that could be activated if and when the need arose.

Apart from several legal developments during the years 1993–2006, which reflected struggles over the geographical boundaries of the salar's fiscal reserve, not much changed in the history of lithium in Bolivia during this period, when the country was wracked by other social and resource conflicts, especially around the gas industry (see above). By the time Evo Morales was inaugurated in January 2006, after the historic elections of 2005, the Chilean lithium industry was already ten years old and growing rapidly; Tesla was about to announce the prototype for its first model (the Roadster); and the Intergovernmental Panel on Climate Change (IPCC) was finalizing its Fourth Assessment Report, which would emphasize the critical importance of global EV adoption as a pillar of climate change mitigation.

This chapter has served several purposes, both theoretical and contextual. Theoretically, it has made an argument about the importance of understanding resource imaginaries, especially those embedded in the green energy transition, as much through their prelives, their multiple prefigurations, as through their afterlives—that is, what remains of the possibilities for social life and meaning-making in the "socio-material ruination of achieved projects of modernization."[75] But in tracing the prelives of lithium through both the different material iterations of lithium, and, more generally, through the histories of resource geopolitics in which Bolivia's unfolding lithium project must be located, I am also making another argument. In looking backward as much as forward, in projecting the critical understanding of lithium in both directions, I am also insisting on a particular approach to the spatio-temporality of extractivism itself.

This is not, however, a way of drawing a hard and fast distinction between extractive projects—especially those involving energy—that are in a state of material and socioeconomic ruination and those that are frozen in a state of perpetual potentiality. Instead, as the study of Bolivia's shifting and multiscalar lithium ambitions reveals, ruination is already present in potentiality at the same time that "achieved projects," whether marked by "decomposition and decay" or not, never manage to shed their "remembered anticipation[s] of the future."[76] As Andrea Ballestero has argued,[77] to study the "future history" of critical resources like water or lithium is to confront a number of ethnographic paradoxes, including the ways in which the everyday lives of deterioration, depletion, and "rubble"[78] are often experienced by people in part through—or even *as*—promises of material transformation and moral renewal.[79]

Beyond the ways in which the past and the future coconstitute the unfolding present of Bolivia's lithium project, the chapter has also underscored the perhaps surprising importance of distinguishing between extraction as a mode of production and the political and economic structures through which extraction is given violent and lasting force. Put another way, we must resist the urge to project particular categories of political or material resistance onto our ethnographic analyses, despite the fact that resistance itself in one form or another is a constant variable. Even more, the enduring tension between extraction, on the one hand, and the terms of entanglement, on the other, must be teased out at quite granular levels.

For example, as the geographer Penelope Anthias has shown, even on what she calls "Bolivia's hydrocarbon-conservation frontier" where communities resist the expansion of state-controlled gas exploration, they do so in complicated ways that belie the often reductive narratives of global environmental activism.[80] Although their different forms of opposition respond to the effects of the climate crisis in Bolivia, including the intensification and impact of seasonal landslides, droughts, and megafires, communities also resist the imposition of overly restrictive conservation regulations by the state, which can have the effect of limiting their room for maneuver, their capacity to put into practice alternative energy futures.

Finally, in tracing the material and geopolitical prelives of lithium up to the beginning of what might be understood as ethnographic time, the chapter has set the stage for a different orientation to follow. Subsequent chapters focus on the people, institutions, and imaginaries that make the fraught struggle to industrialize Bolivia's lithium reserves so revealing about the challenges, even impossibility,[81] of realizing something approximating "climate justice" in an Anthropocene marked by a shared planetary destiny, on the hand, and on the other, by the "savage sorting"[82] of inequality, dispossession, and denial.

TWO

———

"The Fuel That Will Power the World"

IN JULY 2005, I RETURNED to Bolivia during what turned out to be the waning months of the country's neoliberal era, a period of economic, social, and political change that transformed the country in the twenty years after the restoration of democracy in 1982. From the hyperinflation of the early to mid-1980s to the liberal multicultural policies of the 1990s, these two decades would become an iconic epoch with symbolically neat-and-tidy chronological boundaries, an epoch that would later be codified in the radical 2009 national constitution along with the two others—the colonial and the republican—that were seen to precede the historic rupture of 2005, the year Evo Morales and the Movement to Socialism (MAS) party swept into power.

On the eve of this rupture, the cities of La Paz and El Alto had been wracked by months of social mobilization and political upheaval over an interconnected set of long-standing grievances, including the ongoing conflict over the control and commercialization of the country's oil and gas resources. The trauma over the 2003 Gas War, in which dozens of protestors had been killed by state security forces, had not subsided; instead, it remained an ever-present source of searing resentment against the entire structure and logic of Bolivia's national government, regardless of which president was in office.

At the moment of this national transition, the politics of identity and political economy related to each other in complicated ways. Indigenous and "Indianist" frames of reference shaped the discourses of political and social leaders like Abel Mamani, secretary general of the Federación de Juntas Vecinales de El Alto (FEJUVE), Roberto de la Cruz, secretary general of the Central Obrera Regional de El Alto, Felipe Quispe ("El Mallku"), head of the Movimiento Indígena Pachakuti, and the influential activists of the

FIGURE 6. View from an industrial building on the Salar de Uyuni, 2019. Photo by author.

Confederación Nacional de Mujeres Campesinas Indígenas Originarias de Bolivia, known more widely as the Bartolinas. By 2005, the International Labor Organization's Convention 169, the international Bill of Rights for the world's Indigenous peoples, was well established in both international and national law, especially throughout Latin America (Bolivia ratified ILO 169 in 1991). At the same time, as Ronald Niezen has argued, what he describes as "indigenism" had become by the early 2000s a powerful rhetoric of empowerment, a rhetoric with close ties to academic theorizing around the "struggle for recognition" in a "post-socialist age."[1]

But even if indigeneity had itself become a key social and political category in Bolivia by 2005, one distinct from preexisting and more historically grounded ethnic identifications (e.g., Quechua, Aymara, Guaraní), and even if this category had clearly become the centerpiece of a new "moral grammar of social conflict" in Bolivia,[2] it sat uneasily alongside others, including those anchored in the historical materialism of the country's traditional labor unions. Indeed, my own research in July and August of 2005 captured these tensions. I spent most of this time with young Aymara rappers in El Alto,[3] especially with the members of the group Wayna Rap, whose lyrics and artistic self-understanding were part of what the journalist Héctor Tobar described as the "crucible of Bolivia's Indian uprising."[4] But even the members of Wayna Rap talked mostly about resources rather than the politics of identity when describing the causes of the recent mobilizations, which led to the resignation of caretaker president Carlos Mesa and put in motion the electoral process that would culminate in the election of Evo Morales in December.

Yet perhaps in a very early sign of what was to come for future MAS governments over the coming years, its electoral platform during the Second Gas War, which began with blockades and street protests in May 2005, shifted in important ways in only a few months. Unlike the much more radical position of Felipe Quispe and the Movimiento Indígena Pachakuti, Morales and the MAS did not initially demand the full nationalization of the country's hydrocarbon sector. Instead, it had proposed raising taxes and royalty rates on foreign fossil fuel companies working in Bolivia, while leaving the existing system of concessions in place, a system in which the open veins still bleed, just not as much. But by June 2005, with his national political stature growing rapidly and the Mesa government fatally weakened, Morales and the MAS embraced calls for full nationalization, a move that proved to be the main catalyst in propelling MAS to its landmark victory. As will be seen at different places throughout the book, this willingness to change economic strategies in practice while maintaining a discursive fealty to well-established ideological principles would become a hallmark of the country's approach to lithium industrialization.

Despite the fact that resource politics more than a politics of identity were at the center—or, rather, the base—of the historic election of Morales and the MAS, lithium itself was completely absent from the demands of this resource politics, at least at the national level. Even though the regional conflicts of the early 1990s over lithium described in the last chapter remained in the collective (regional) memory, as will be seen, this legacy did not play a measurable role in the political economic struggles that brought Morales and the MAS to power. In many ways, the status of the Salar de Uyuni in the national consciousness, like the status of Potosí Department, remained very much on the margins.

Notwithstanding its important—if, I would argue, superstructural—gestures toward the cultural consequences of its electoral triumph, including the creation of a vice-ministry charged with implementing a state policy of decolonization, the new MAS government was largely preoccupied with the question of how to finance its ambitions for what came to be known as the "process of change." It would turn its attention almost immediately to gas and oil, especially gas; still, the state's much more fraught lithium project was launched not long after, a project that slowly unfolded at the margins of the country's gas "bonanza." The lithium project would contribute very little to the country's economic growth from 2006 to 2020 (the timespan examined in this chapter), a period during which Bolivia became an international poster child for macroeconomic stability (with its World Bank-and-IMF-pleasing levels of foreign exchange reserves as a percentage of its GDP), but its actual importance during these years must be understood in different terms.

Even though the MAS government was quite happy to depend on the truly transformative economic benefits of the "Age of Oil" while it lasted,[5] this dependence was recognized as problematic—on environmental, economic, and ideological grounds—from the early years of the first Morales administration (2006–2009). Natural gas and oil are resources of the past, despite their ongoing critical importance both to countries (like Bolivia) that rely on their extraction and to the wider global geopolitical system whose history is partly explained by the widespread replacement of coal with oil. But as will be seen, lithium was always associated in Bolivia with the future, an alternative *energy* future, one in which a resource with strategic national importance and value would also play a central part in solving global environmental problems for which Bolivians themselves were not responsible.

In this sense, Bolivia's lithium, its potentiality, was conceived by policymakers as a gift to the world, with profound economic and moral dimensions. The country's massive lithium reserves might one day replace gas as the economic foundation of its multigenerational process of change, its "democratic and cultural revolution." But even more, the offering of Bolivia's lithium to the global energy transition was seen as a moral gesture of historic dimensions. The country that was exploited for so many centuries, the country whose plunder made the rise of global capitalism possible, would nevertheless give to the world a resource that would prevent it from dying, from suffocating on its own carbon emissions, a resource that would allow Bolivia, once and for all, to "leave in the past" (as the 2009 constitution puts it) its "colonial, republican, and neoliberal" histories.

This chapter examines the trajectory of lithium's material, political, and moral lives in Bolivia during two distinct, but interconnected, periods: first, the thirteen years of governance, conflict, and economic transformation during the different Morales administrations (2006–2019); and second, the traumatic year of the coup d'état, right-wing revanchism, and the pandemic (2019–2020). These two distinct periods coincided with wholesale changes in the broader global status of lithium, changes that shaped resource politics in Bolivia and were, in important ways, shaped by them.

The next section of the chapter considers the history and ethnography of the MAS turn toward lithium. In many ways, the MAS government brought an unmistakable vision to at least the idea of lithium industrialization, even if the different steps it took to realize this vision were delayed, thwarted, or poorly conceived. But in other ways, the evolution of the lithium project during the period 2006–2019 was shaped by forces that had nothing to do with lithium potentiality at all.

On the one hand, the Morales government and its institutions were buffeted by political and social conflicts—some of which were clearly existential—almost without pause from his inauguration in January 2006 to his ouster in November 2019. If the lithium project emerged in the shadows during this same period, these were shadows cast by a series of crises that threatened not just the ambitions of the historic Morales presidency but also the viability of the wider process of change for which the MAS government was meant to be the political instrument.[6]

But on the other hand, even with global demand for lithium increasing dramatically, especially during the third Morales administration (2015–2019), the government continued to rely on revenues from hydrocarbons, at the same time as it pursued a number of other economic development projects—for example, the industrial production of urea for fertilizer. Yet given the range of technological and political questions that had always swirled around the industrialization of lithium and other "evaporitic resources" at the Salar de Uyuni, the MAS governments under Morales were not able—or willing—to mount the kind of national socioeconomic campaign that would have been necessary to confront and resolve these questions.

The chapter then turns to an extraordinary year of structural violence and a public health emergency in Bolivia and the multifaceted ways in which the different shocks during this year left lasting marks on the ever-shifting lithium project. The coup d'état that unfolded in queasy slow-motion over these many months must be understood in relation to longer political and cultural histories—a history in which the ascent to power of a self-identifying Indigenous president would have been unthinkable and one in which this same president would turn his back on cultural norms in seeking to extend his power. But what role did lithium play in the more immediate circumstances that saw Morales and other high-ranking MAS officials forced into exile? Despite the fact that the ouster was widely described—especially by Morales himself—as the "Lithium Coup," what does the ethnographic and other kinds of evidence reveal about this way of characterizing this *annus horribilis*?

The chapter concludes by reflecting more generally on how the fourteen years between the inauguration of Evo Morales in 2006 and the reelection of a MAS government in late 2020 were marked by a pervasive tension between the allure of lithium industrialization and the peril of its unrealized promise of national economic and social transformation. Drawing from research on other cases in which countries found themselves trapped in the elusive webs of resource potentiality, I argue that Bolivian policymakers and local communities alike found themselves lost in the disorienting maze of "first lithium," a condition of endemic indeterminacy that fueled social and political conflict.

FROM POLITICAL TRANSFORMATION TO PRODUCTIVE SOVEREIGNTY

The Morales years in Bolivia began amid symbolically charged pomp and circumstance, including an unprecedented inauguration ceremony at the pre-Columbian ruins at Tiwanaku, where the new president received the blessings of a group of prominent *yatiris*, or Aymara ritual specialists, arrayed on the steps of the Kalasasaya Temple. Very soon after, the new MAS government returned to the pressing question of resource development. On International Workers' Day 2006 (May 1), Morales ordered the nation's armed forces to occupy the country's oil and gas fields, which were largely under the control of private foreign companies like Brazil's Petrobras and Spain's Repsol. This old school nationalization sent shock waves across regional markets and political alliances, despite the fact that Brazil's president at the time, Luiz Inácio Lula da Silva, or simply Lula, was a strong supporter of Morales and a longtime advocate for labor unions (though not necessarily state socialism).

As an article in the *New York Times* put it, "many industry observers feared [the nationalization] would scare away investors and jeopardize regional economies." The article quotes from an energy analyst, who makes a comparison with Saudi Arabia, one that would also be made repeatedly in the international media in relation to lithium: "[Bolivia] isn't like Saudi Arabia, which over the years has developed a know-how to dominate the [oil] industry independently."[7]

But while the MAS government focused on managing the international blowback to hydrocarbon nationalization, at the same time it confronted both a growing separatist movement centered in the city of Santa Cruz and antigovernment mobilizations in the historic city of Sucre. Meanwhile, local and regional actors in Potosí Department were moving forward independently in an attempt to spur lithium development.[8] In 2007, the Federación Regional Única de Trabajadores Campesinos del Altiplano Sur (FRUTCAS)—the influential federation of peasant-workers we encountered in the last chapter—submitted a detailed plan to the MAS government, which FRUTCAS had overwhelmingly supported in the 2005 elections, for the industrialization of both lithium and potassium at the Salar de Uyuni.

The technical aspects of the FRUTCAS proposal had been designed by a mysterious Belgian borax entrepreneur named Guillaume (aka Guillermo) Roelants du Vivier, who had spent several months in a Bolivian prison after being arrested in 2000 by a combined US Drug Enforcement Agency (DEA), Bolivian, and Chilean task force. He had been accused of using his factory at Apacheta on the Bolivian-Chilean border to launder large amounts of sulfuric acid. Although used in the production of boric acid, Roelants du Vivier couldn't explain why his factory had much more sulfuric acid—which is also used to process coca paste and thus cocaine—then it could possibly use for legitimate industrial purposes.[9] By 2007, however, Roelants du Vivier had reemerged in the town of Río Grande near the Salar de Uyuni, where he partnered with community members in establishing a local mining company called Sociedad Colectiva Minera Río Grande, which produced the borax by-product ulexite, used in paper products and fiberglass.[10]

The FRUTCAS proposal for lithium industrialization was anchored in the principle of state control, something that distinguished it from the flourishing Chilean lithium industry, which was dominated at the time by the private SQM, the Pinochet-linked mining company that had "been known to bankroll Chile's political scene, both the left and the right."[11] Given the abstractness of the potential economic consequences of the lithium project, FRUTCAS's bold move must be seen in light of three factors: first, its desire to deepen its political ties with the national MAS government at a historic moment of sociopolitical transition and liminal possibility; second, its interest in making clear that it would defend the rights of local communities in a future regime of royalty distribution; and third, and perhaps most important, its willingness to distance the federation—which represents dozens of communities across five rural provinces—from the claims of COMCIPO, the civic organization based in the departmental capital of Potosí.

This cleavage between FRUTCAS—and several other, less influential local and regional movements—and the social and political institutions in the city of Potosí would remain a constant source of tension at the heart of the national lithium project in Bolivia. Yet it was a tension that could at times actually work to the project's advantage, especially given the fact that FRUTCAS—whose member communities were closest to the salar—had been a reliable base of support for the national MAS government, while COMCIPO (despite its place within a wider pro-MAS departmental ecosystem) emerged as one of the main, and most surprising, voices of anti-MAS dissent, an orientation that would culminate in its highly visible support for the 2019 coup (see below).

Nevertheless, the MAS government's response to the lithium proposal from below showed the limits to both FRUTCAS's own ambitions and to local and regional power in relation to extractive projects that are considered pillars of the country's economy. In April 2008, Morales issued Supreme Decree 29496, which put the industrialization of the Salar de Uyuni under the control of COMIBOL, the Bolivian state mining company. It also declared lithium a "national priority" and authorized up to $5.7 million of state funding to build the initial infrastructure, including evaporation ponds, wells, pumping facilities, and a small "pilot plant" for the production of lithium carbonate.

Yet at the same time, in a development that antagonized FRUTCAS, the government opened up negotiations over lithium industrialization with a number of foreign companies and countries, including Mitsubishi, Sumitomo, the French conglomerate Bolloré (which has a subsidiary EV business), the South Korean state company Korea Resources Corporation (KORES), and the governments of Brazil and Iran.[12] These negotiations provoked FRUTCAS to issue a carefully worded resolution that ratified its "total support for the initiative of President Evo Morales" to move forward with a lithium project that was "100% state controlled," while reiterating that the project ultimately belonged to the people and communities of the salar region, who were the project's "guardians and monitors." The resolution concludes with the exhortation, "No to the privatized exploitation of the Salar! No to international tenders!"[13]

However, even as COMIBOL broke ground for its pilot plant on the edge of the salar, at a site known as Llipi, the national MAS government—and the country itself—was in the grip of a series of crises that brought Bolivia, as I have argued elsewhere to the brink of civil war.[14] This period of crisis opened up on two distinct but ideologically commensurate fronts. On the first, the historically oppositional civic movement in Santa Cruz, under the banner of "Camba" ethnonationalism, launched a separatist campaign that included the takeover of government buildings and the stockpiling of weapons for use in the event of what militants expected would be an armed rebellion against the "dictatorial" MAS regime in La Paz.[15] And on the second front, the city of Sucre mobilized against the Constituent Assembly, a body that had been convened in the legal capital of Bolivia to write a new national constitution.

This period of national trauma and violence, which, among other things, rendered a more concerted focus on the incipient lithium project impossible, was marked by two signal moments: the capture and humiliation of Indigenous and peasant MAS supporters by an urban mob in the historic central plaza of Sucre, a day of atrocity that was livestreamed to the horror of audiences throughout Bolivia; and, in September 2008, the massacre of MAS supporters near Porvenir in Beni Department by armed groups acting at the direction of the department's governor.

At the same time, throughout 2008, especially during the last months, Morales was engaged in a national political campaign on behalf of the new constitution, which would be put to a vote in January 2009. Even more than an earlier national vote of no confidence, which the MAS government had easily survived in August 2008, the vote on the constitution was seen as an epochal moment for the country itself, a choice between the revolutionary vision of Morales and the MAS, including its plan for state-led economic development, and the separatist visions of the various anti-MAS opposition movements. In a process of compulsory voting, the new constitution was approved by over 61 percent of Bolivians along regional-political lines, "refounding" Bolivia as a plurinational state. Importantly, Article 369 of the 2009 constitution enshrined into law state ownership of the country's "mineral riches" and declares that the "non-metallic natural resources found in the salares [and] brines" are of strategic national importance.[16]

The years after the vote on the new constitution, which was a turning point after years of sociopolitical turmoil and mobilization, both revolutionary and separatist, resembled a version of the Thermidorian Reaction, except that it was the MAS government itself that turned increasingly away from the more radical dimensions of the process of change to focus on the bureaucratic and pragmatic realities of governance. It was also able to finally return to the lithium project, which had—from its isolation at the salar—continued to move forward, mostly through the construction of the pilot plant at Llipi. However, beginning in 2010, the government began to raise the stakes of the lithium project significantly.

In order to give institutional form to the strategic importance of lithium, the government created the Gerencia Nacional de Recursos Evaporíticos (GNRE), a new entity still nominally under the direction of COMIBOL but with much greater independence. Then, several months later, in November 2010, the new Plurinational Assembly authorized $120 million in funding to GNRE specifically for the industrialization of lithium. A year later, in December 2011, a second credit was issued to GNRE by the Central Bank of Bolivia, this time for $765 million, bringing the total state investment in the lithium project to almost $900 million.[17]

Yet even though this massive second round of state financing was directed to GNRE, the actual law authorizing the credit left the political economic door open to more flexible strategies of industrial development. Although the money could be used by GNRE as "direct capital investment," it could also be used to cofinance joint ventures with "technology companies" who would want to invest in knowledge transfer or even production along the entire "evaporitic chain" (Law 211, Art. 33). Despite the long-standing opposition in the salar region to the involvement of private foreign companies—"No to the privatized exploitation of the Salar! No to international tenders!"—the funding law provided no further details about, or limitations on, the nature of the technology companies with whom GNRE might partner.

Throughout 2012, the Morales government—having been reelected, again by a historically wide margin in December 2009—was at work formulating a doctrine to support the wider process of change, one that would tie together the various, and often disparate, threads in what was an evolving ideology of economic, social, and political transformation. By January 2013, the doctrine was ready. Not surprisingly, it was framed in relation to the country's bicentennial in 2025 (Bolivia declared its independence from Spain on August 6, 1825), a time horizon that gave a strong indication of Morales's intention to lead the process of change over the long-term, despite existing constitutional term limits.

The doctrine was called the "Patriot Agenda 2025" (PA 25) and it announced "13 pillars for a dignified and sovereign Bolivia."[18] In many ways, PA 25 represents the culmination of MAS's political and ideological development, the clearest statement of its historical ambitions, expressed in terms that sought to interweave political economy, revolutionary praxis, and a politics of liberation into a vision that was both distinctly Bolivian and directed toward wider regional and global audiences. It is a doctrine for the epoch that comes after a number of interconnected chains are broken, including the chains of "colonial capitalism, liberalism, and the neoliberalism of the *patrones*"—that is, the private bosses and owners of "land, of mines, of hydrocarbons, and all our natural resources." As the preamble to PA 25 explains, the process of breaking these chains must be seen more generally as a consequence of the Pachakuti, the millennial rotation of space-time that gives way to an enduring era of widespread "equilibrium" beyond the constraints of human, and especially capitalist, temporalities.[19]

Of the thirteen pillars of PA 25, pillar 6 would come to have the most lasting discursive impact on Bolivia's lithium project; it would continue to shape the country's planning well beyond the end of the Morales period itself. Pillar 6 represents an innovative intervention, in which "production" and "productivity" are reconceptualized beyond what it calls the "dictatorship of the capitalist market." Pillar 6 distinguishes between forms of *production*, which are simple mechanisms of capitalist exploitation—in which, as we have seen, Bolivia has played an indelible role throughout global economic history—and *productivity*, which pillar 6 reimagines as a sociopolitical value. Even more, as a sociopolitical value, productivity is understood as a necessary—though not sufficient—condition for the realization of a lasting and even emancipatory sovereignty beyond the boundaries of the country's "colonial and republican heritage."

But despite this reformulation, "productive sovereignty" is not in tension with either industrialization or the extraction of the country's most valuable resources, even as measured by the logics of the same global capitalist markets against which pillar 6 is meant to serve as an alternative framework for economic development. Indeed, quite the contrary. According to PA 25, productive sovereignty can only be fully achieved when the country's most critical resources, including lithium, are "consolidated" through processes of industrialization that manage to expand productive capacity while repairing and remediating the "zones of life" that will necessarily be damaged despite the use by the state of the "best technologies available."

Perhaps most strikingly, the realization of productive sovereignty is not incompatible with the insertion of Bolivia's resources into global capitalist value chains; the critical element, as we have seen, are the terms of entanglement rather than the fact of entanglement itself: "*To the maximum extent possible*, and without creating either dependency or submission, Bolivia will take advantage of the benefits of commercial agreements in order to promote opportunities to export the products produced in the country" (PA 25, pillar 6, emphasis added).

It is important to understand that productive sovereignty—which would become the doctrinal basis for Bolivia's lithium project—is not synonymous with more strategic claims for *resource* sovereignty. Within the wider constellation of what John-Andrew McNeish has called "sovereign forces" in Latin America—that is, the processes through which ambitious visions for social and political change must grapple with environmental and economic constraints—"productive sovereignty" must not be seen as simply a proprietary claim over the nation's resources.[20] Rather, it represents a much more far-reaching attempt to reimagine *both* the grounds of the country's political economy and its framework for political legitimacy. In other words, productivity is reconceptualized as an instrument of liberation, while sovereignty depends on this instrument being given a fully realized shape and form.

Although it has become almost a truism within the critical literature on Bolivia to argue that the revolutionary ambitions of the MAS governments, especially under Morales, were eventually abandoned—some faster than others—in the face of different factors, from the threat of right-wing antigovernment mobilization to the increasing authoritarianism of the MAS political elite, the relationship between the doctrine of productive sovereignty and the lithium project reveals something else. As will be seen, productive sovereignty remained ideologically capacious enough to explain and justify a wide range of developments over the years. But everything that happened—and didn't happen—after 2013 was already prefigured in the text of PA 25: the focus on lithium *industrialization*; the way the lithium project was promoted in political, as much as in economic, terms; the willingness to accept the prospect of some environmental damage as a necessary consequence of lithium extraction; and, perhaps most importantly, the fact that a reliance on "commercial agreements" with foreign companies and states became itself a pillar of the lithium project, not as a compromise with the ideological principle of productive sovereignty, but as a practical expression of it.

PRODUCTIVE REALITIES AND CONSTITUTIONAL CONFLICT

Less than a year after PA 25 was promulgated, Bolivia held another national election in which Morales and the MAS were returned to power again by a historic margin (61%), although less than in 2009 (64%). With the process of change ratified yet again after almost a decade of conflict and transformation, the Morales government put the tenets of productive sovereignty into dramatic action. Although GNRE was producing around 1.5 tons of lithium carbonate per month at the pilot facilities of Planta Llipi by 2015, the technicians had only managed to achieve a "technical grade" (also known as "commercial grade") product—lithium carbonate that has a use in applications like ceramics or industrial lubricants but not in lithium-ion batteries. But to produce even technical grade lithium carbonate at a pilot level, GNRE had to complete much of the infrastructure in the productive process, including building the brine evaporation ponds, installing the pumping system, and establishing the technical capacity among GNRE workers to purify residual salts through chemical intervention and analysis.

With most of the important parts of this proof-of-concept model in place, the MAS government made the historic decision to move from the small-scale and experimental phase in the lithium project to the industrial. Still, with questions still unanswered about GNRE's ability to produce battery grade lithium carbonate even at a pilot level, the government opted for a compromise solution. Instead of moving forward immediately with plans to produce lithium carbonate at an industrial scale, it decided to apply the lessons from the proof-of-concept strategy to the industrialization of something else: potassium chloride.

As GNRE technicians at the salar realized very early in the process, the production of commercial potassium chloride, which is used as fertilizer in large-scale agriculture, is much less demanding than the production of lithium carbonate. Although the process is the same through many of the stages—from brine evaporation to the production of commercial salts—the level of purity and chemical composition required are very different. Commercial potassium chloride contains around 50–52 percent potassium and 45–47 percent chloride, while battery grade lithium carbonate must have a purity of at least 99.5 percent. At the same time, as will be seen in more detail in the next chapter, the brines of the Salar de Uyuni are also rich in magnesium, another alkali metal whose chemical similarities with lithium pose significant problems during the final stages of producing lithium carbonate; they do not pose the same challenges in the production of potassium chloride.

In July 2015, the MAS government signed a landmark deal with CAMC Engineering Co., a $17 billion Chinese company with headquarters in

Shenzhen that manages hundreds of infrastructure and building projects throughout Asia, Africa, Latin America, and Eastern Europe, which makes it one of the main actors in China's Belt and Road Initiative. The $178 million contract between GNRE and CAMC was for the company to build what would be one of the ten largest potassium chloride plants in the world and one of the largest industrial facilities of any kind in Latin America.[21] Although CAMC would be responsible for all phases of construction and delivery, GNRE had purchased the architectural designs for the 2,500-square-meter plant from the German company ERCOSPLAN, based in Erfurt, which had decades of experience in the potash industry. Its company motto was the enthusiastic "We do everything about salt!" This model was followed a month later, in August 2015, by a second contract. GNRE hired another German company, K-UTEC, to begin a ten-month study for a *future* industrial lithium plant, despite the various doubts about the feasibility of actually producing lithium carbonate from Bolivian brine.

With this first phase in the long-term state-led lithium project more or less completed, and despite with the technical setbacks that compelled the government to begin with the industrialization of potassium chloride rather than lithium, the Morales administration and GNRE officials were becoming increasingly confident in the project—and its symbolic importance. As Morales put it, during the ceremony in which the design contract with K-UTEC was signed, "These investments are our own money. We will have no partners. We will be the owners." Luis Alberto Echazú, the director of GNRE, added: "We can say with much pride that everything we have accomplished up until now has been through our own efforts, our own funding, our own investigations, indeed, through the efforts of all Bolivians."[22] And in early 2016, during one of his many policy speeches, Álvaro García Linera, the country's vice president, went even further. Even though the actual prospects for lithium industrialization remained tenuous at best, the MAS government was certain about one thing: Bolivian lithium would one day be the "fuel that will power the world."

Yet while construction of the potassium chloride plant got underway at the salar, the institutional dimensions of the lithium project underwent another important change: GNRE, which had been subsumed within COMIBOL, was replaced with Bolivia's new state-owned mining company, Yacimientos de Litio Bolivianos. The April 2017 law authorizing the creation of YLB gave the state company responsibility over the entire "productive chain," from exploration to commercialization, while drawing a key political economic distinction—one consistent with the doctrine of productive sovereignty (see above)—between the preindustrial and industrial links in this chain. Although YLB must guarantee "100%" state control over all the "basic chemical processes" required to extract and refine the materials needed to produce lithium carbonate, "subsequent processes of semi-industrialization, industrialization, and waste processing" can be carried out through joint ventures with "national or foreign private companies," as long as YLB retains at least a 51 percent ownership stake.

In the meantime, however, the country was facing a constitutional conflict of the MAS government's own making. With the 2019 general election looming just over the horizon, the Morales government took the fateful decision to petition the Plurinational Constitutional Court to abolish the term limits provision of the country's constitution, which would allow Morales to run for a third term in office. Although critics of Morales—both within and outside of Bolivia—would read the court's subsequent November 2017 ruling overturning term limits as a sham, given the fact that most of the justices were affiliated in different ways with MAS, the actual legal arguments were quite nuanced.[23]

Regardless of these complexities, the political costs of the court ruling to Morales and the wider process of change were enormous. As Linda Farthing and Thomas Becker explain, the November 2017 decision

> handed Morales's opponents the very tool they had lacked. Even though these adversaries were fractured by constant infighting, they were buoyed by the onslaught of negative public sentiment toward Morales and managed to cobble together an opposition. Known as 21F, the coalition spanned the gamut from traditional right-wing parties, to regions convinced the Morales government had shortchanged them, to environmentalists angered by the government's extraction-heavy model.[24]

Although this tool would soon be wielded in an attempt to undermine the lithium project, among other things, the following year was actually marked by a number of major advances for YLB. In October 2018, in a vibrant ceremony on the salar attended by Morales, García Linera, government ministers, YLB technicians, and local authorities, including those from FRUTCAS, the industrial potassium plant was inaugurated, with a whopping capacity to produce 350,000 tons of potassium chloride per year. In his speech at the event, Morales made a direct link between the new plant, the wider lithium project, and the MAS government's system of redistributing earnings from state industries as social benefits, arguing that future lithium industrialization would allow the country to "continue improving the benefits for our people, especially for our children and grandparents."[25]

Also in 2018, after an accelerated public tender and process of evaluation, YLB signed a $96 million contract with the Chinese companies Maison Engineering and China Machinery Engineering Corporation (CMEC) to build the industrial lithium plant based on a design submitted by the German company K-UTEC. The contract called for the plant to be completed within fourteen months with a capacity to produce fifteen thousand tons of lithium per year. But while obviously important, the deal with Maison Engineering and CMEC was overshadowed by a parallel negotiation process, one that would play a key role in the historic crisis of 2019–2020.

BREAKING THE "CHINESE MONOPOLY"

At the same time as Bolivia's close industrial and infrastructural ties with state-backed Chinese companies rightly drew much of the critical attention—and still does—the MAS government was being courted by the German technology sector. This concerted commercial diplomacy didn't take place in a vacuum; the major German technology conglomerate Siemens had already built three power plants in Bolivia as part of the rapid expansion of the country's gas industry.[26] But the intense German interest in forging a major partnership with YLB and Bolivia was driven by very different factors. By 2018, the negotiations that would culminate in the €1 trillion European Green Deal (EGD) were already well underway within the European Commission, even though the decision wouldn't be officially announced until January 2020 (see the introduction). As we have seen, the launch of the EGD sent ripples through the entire global lithium energy assemblage, ripples that were felt as spikes in the price of lithium, disruptions within battery supply chains, and exponentially rising valuations of companies like Tesla.

Given the fundamental importance of automobile manufacturing in particular to the German economy and its sociopolitical identity, the impending surge in demand for lithium-ion batteries for EVs exerted a unique pressure on the country with the largest economy in the EU.[27] In an effort to break the dependence by German—and, more widely, European—automakers on South Korean, Japanese, and especially Chinese battery producers, both the German government and private companies sought to secure long-term access to the raw materials needed to build battery capacity within Europe

FIGURE 7. Chinese building contractor claims it will "create a more prosperous tomorrow," Salar de Uyuni, 2020. Photo by author.

itself. Among these different raw materials, as we have seen, lithium is considered the most critical. Moreover, given the fact that Bolivia's lithium reserves are both the largest in the world and still largely locked in the brines of the country's salares, German politicians and companies viewed a deal with YLB to be of pressing national importance.

As a response, "Berlin led an intense lobbying effort [with Bolivia] on a political level. It included several diplomatic visits during which the advantages of moving forward with German companies was pitched."[28] In addition, "Bolivia's officials were . . . invited to tour German factories, in order to experience German industrial prowess first-hand."[29] This lobbying concluded with a personal letter from Peter Altmaier, the German Minister for Economic Affairs and Energy, to Morales himself, in which Altmaier argued that German industry is committed to economic development that is consistent with Bolivia's state ideology of *vivir bien*, or "living well."[30]

Despite the fact that Chinese companies were also working to strike a deal with YLB beyond the relatively modest construction contracts for the potassium and lithium industrial plants, the German politico-technological charm offensive made the difference: in December 2018, during ceremonial events in both La Paz and Berlin, YLB signed a contract for what promised to be one of the most ambitious, and riskiest, projects in Bolivian economic history. Even more stunning was the identity of the German firm with whom YLB formed a landmark joint venture: ACI Systems, a relatively unknown company with no discernable history or technological expertise in the kind of wide-ranging processes at the heart of the deal.

The story of ACI Systems begins with Wolfgang Schmutz, a mechanical engineer, inventor, and business developer from Zimmern ob Rottweil. After graduating from the University of Stuttgart in 1980, he worked for ten years in research and development for the Fraunhofer Society of Manufacturing Engineering and Automation, where he was part of the team that invented a residue-free cleaning system that uses CO_2 ice crystals to bond with and remove microparticles from a variety of materials during the production of, for example, electronics and computer motherboards, medical imaging machines, and most importantly, various automobile components.

In 1997, several colleagues from Frauhofer founded a spin-off technology company called acp systems AG in order to market the CO_2 dry cleaning system and develop its different applications. Over the subsequent years acp systems developed a number of patents for more specialized or portable versions of the original system and eventually built a small assembly plant on the outskirts of Zimmern ob Rottweil and an R & D facility in the town of Ditzingen, a suburb of Stuttgart. In the meantime, Schmutz had also started another company, located at the same location in Zimmern ob Rottweil, called ACI Group, which focused on the production and distribution of solar panels.

It was this dual institutional history that partly explains the mystery of how a small "clean tech" and solar technology company from southwestern Germany found itself at the center of negotiations with YLB over a wildly ambitious new phase in Bolivia's lithium project.[31] With his close ties to the German automobile sector, which uses the CO_2 cleaning system in their manufacturing plants, Schmutz was able to make the argument that his small team from Zimmern ob Rottweil could develop the technology that would connect German car companies directly with the largest source of lithium in the world just at the moment in which the EV transition was becoming a major economic and political imperative for Germany and the EU.

Indeed, although the December 2018 agreement, which was followed by a Bolivian presidential decree (SD 3738) that officially created the joint venture, called "YLB-ACISA" (for "ACI Systems Alemania"), was vague about certain parts of the seventy-year project, the central objectives of the deal were clear enough: ACI Systems would develop the technology and industrial capacity to produce battery grade lithium *hydroxide* (not carbonate) from residual brine, which is the brine that contains lithium (and other minerals, like magnesium) that is not "recovered" after the initial process of brine evaporation is finished.

This proposal had two important dimensions, at least on paper. First, it would have allowed YLB to finish and launch its industrial lithium carbonate plant as planned, while adding to the wider industrialization strategy by potentially increasing the overall production of lithium carbonate from fifteen thousand tons to at least thirty thousand tons per year (which is what the YLB-ACISA agreement projected). And second, the deal reflected sensitivity to the fact that by 2018, battery manufacturers were increasingly showing a preference for lithium hydroxide, which must be made from lithium carbonate and is thus more difficult to produce, something indicated by the higher prices per ton for the by-product.

Beyond the main provision to build an industrial lithium hydroxide plant on the Salar de Uyuni, the YLB-ACISA contract also included a final component: guaranteed access to the German market for the battery grade lithium hydroxide. Indeed, it is this provision that goes the furthest in answering many of the questions that were raised about the role of ACI Systems when the deal was announced. As Wolfgang Tiefensee, economic minister of Thuringia and member of the German delegation to Bolivia put it, "This partnership secures lithium supplies *for us* and breaks the Chinese monopoly."[32]

In other words, all the gestures by ACI Systems to the more symbolic dimensions of Bolivia's lithium project aside, the practical meaning of the 2018 agreement from the German side was the protection of national and European economic interests. The deal would have directed *at least 80 percent* of the battery grade lithium hydroxide to German automobile and future battery manufacturers. The effect of an exclusive supply of at least thirty thousand tons per year would have been to reassure the German government, German states (like Thuringia), and private companies that they could go forward with plans to invest the hundreds of billions of euros needed to build a German/European battery sector.

Yet, in the event, nothing turned out as planned—for Bolivia and YLB; for ACI Systems; for investors in Thuringia (the site of the future German battery industry); or for officials at the highest levels in Germany itself, including Peter Altmaier, the German Minister for Economic Affairs and Energy, who had wooed Evo Morales with a personal letter. Although the social and political conflict that followed was the result of a confluence of factors, it began with the structure of the YLB-ACISA agreement itself. Unlike the existing "commercial agreements" with the German and Chinese companies to design and build the industrial potassium chloride and lithium carbonate plants, the YLB-ACISA deal created something quite different: a true joint venture, in which YLB would own 51 percent but a *private foreign company*, ACI Systems, would own the remaining 49 percent.

Thus, while the arrangement might have strictly complied with the mandate of PA 25, which views "the benefits of commercial agreements" as fully consistent with, indeed, an expression of, "productive sovereignty," news of ACI Systems' 49 percent ownership stake sparked significant opposition, especially in the city of Potosí, where the civic movement COMCIPO declared a general strike "in defense of lithium" in August 2019. Two months before the fateful elections in October, the anti-YLB-ACISA mobilizations would set the stage for the coming upheaval.[33]

THE WHITE POWDER AT THE END OF THE WORLD

But the civic activists of COMCIPO, which became a surprising thorn in the side of the national MAS governments, weren't thinking about PA 25 or productive sovereignty during the highly charged period after the mobilization had begun, at least not in the same terms. As a "civic" movement, COMCIPO fights for the interests of the city of Potosí itself, a long-term struggle against what it perceives as a legacy of neglect from the national governments—regardless of political party—in La Paz. However, this enduring sense of economic and political abandonment underwent a profound transformation in the post-2006 period. Unlike the civic movement in Santa Cruz, for example, which is embedded in a wider department that was always hostile to the MAS governments in La Paz, the city of Potosí is the main urban center in a department that was—and remains—strongly pro-MAS. Many COMCIPO members, at least during the first and second Morales governments, sup-

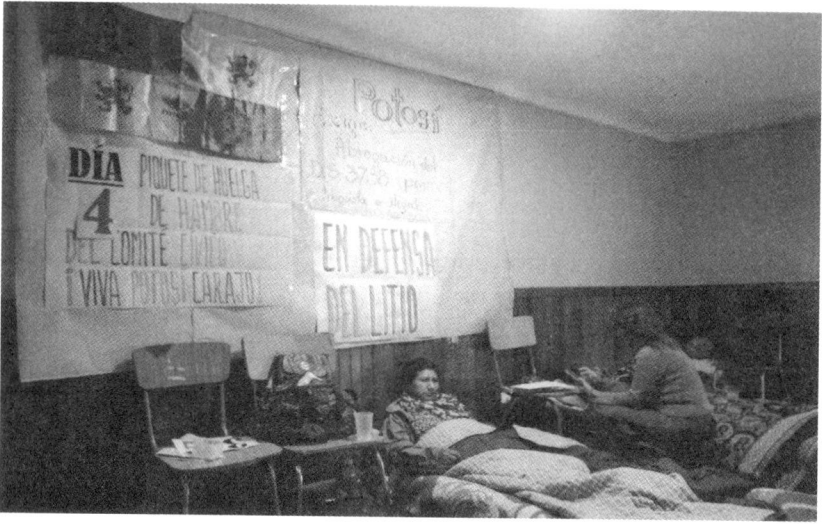

FIGURE 8. Militants from COMCIPO launch a hunger strike "in defense of lithium," La Paz, 2019. Photo courtesy of Miriam Jemio.

ported the country's first self-identifying Indigenous leader, along with MAS's vision for the process of change.

The political orientation of COMCIPO began to shift fairly dramatically from the beginning of Morales's third term in office. The crystallization of an anti-MAS and, even more, anti-Morales, politics in the city of Potosí first exploded into view in July 2015, when hundreds of miners, COMCIPO leaders, and students descended on La Paz after a riveting journey from Potosí, which was covered extensively by the national media.

One of the leaders of the 2015 protest was Marco Pumari from COMCIPO, a charismatic activist who became the voice of *potosino* grievance. And what were they demanding? Instead of waiting patiently in the former "Imperial City" while other cities and departments received largesse from the country's gas bonanza, the activists marched through the streets of La Paz, blockading routes and detonating dynamite, to demand new investments for Potosí, including the construction of a new airport.

Nevertheless, the Morales government was dismissive of the mobilization and dispatched Vice President García Linera to ridicule the civic leaders from Potosí. As he put it in a speech, "To say that nothing has been done [for Potosí], that we have forgotten, is a lie, an injustice, and a fallacy. . . . In reality, Potosí gives very little these days. It used to produce much but now there is much more money to be made through gas and Potosí doesn't produce gas."[34] Smarting from the official rebuke, and having gained nothing of substance for their efforts, the COMCIPO-led activists made the long journey back to Potosí.

Four years later, with the national elections only two months away, and a more global cause for which to mobilize—the "defense of lithium"—COMCIPO was ready. After announcing the general strike in August 2019, COMCIPO raised the stakes by declaring in October that several of its leaders, including Marco Pumari, would go on a hunger strike until the MAS government repealed the YLB-ACISA agreement. In her reporting from Potosí and La Paz during these final weeks before the elections, the Bolivian journalist Miriam Jemio documented just how much the lithium project had become a powerful new plank in COMCIPO's antigovernment platform. She quotes from Pumari, whose status had only risen since the 2015 protests, as he lay weakened by hunger under a handwritten sign that read "Potosí demands the cancellation of the treasonous and illegal SD 3738 [which created YLB-ACISA]": "We have been deceived by industrialization. They have been telling us for more than 10 years that lithium batteries are going to be produced and until now there is nothing."[35]

However, while the protests and hunger strike "in defense of lithium" continued in Potosí and La Paz, the country prepared for what promised, under any circumstances, to be a highly contested national election. As we have seen, support for Morales had been plummeting from its historic highs since the November 2017 constitutional court ruling, at the same time that a new opposition coalition was emerging that interconnected the traditional anti-MAS right wing in Santa Cruz with COMCIPO; a growing anti-Morales left associated with Bolivia's urban universities; and lowland Indigenous movements, which had a history of opposing state-led infrastructure development and extraction projects in their territories. Nevertheless, few could have predicted what happened after voting took place on October 20.

At the end of the day, vote counting was suspended for twenty-four hours for reasons that were unclear at the time, after which the national electoral tribunal declared Morales the outright winner by a slim margin of just less than 11 percent over the next candidate, Carlos Mesa. Apparently, Morales had won in the first round (since the difference between the first and second candidates was more than 10 percent). As Farthing and Becker have shown in their convincing analysis, the newly configured opposition coalition was already prepared.[36]

Using the full arsenal of social media at their disposal, and with the support of a broad international array of anti-Morales actors, the opposition launched or encouraged widespread protests against the elections across the country. Although some of the protestors in cities like La Paz, including those who had been MAS supporters in the past, genuinely thought that electoral fraud had taken place, the assertion that Bolivian citizens rose up en masse against Morales and MAS to save democracy was a key weapon in the targeted campaign of disinformation by the highly organized cabal known as 21F.[37]

As the national political crisis worsened through late October, the Morales government looked for a strategy to blunt the momentum by dividing the opposition. To do so, it turned to the conflict over the YLB-ACISA agreement, thinking that a bold stroke could somehow bring Potosí and COMCIPO onto its side. On November 2, Morales issued a new supreme decree. After "considering" that it was "necessary to make adjustments [from time to time] to the politics of lithium industrialization ... in order to achieve optimal conditions of national and regional scope," the decree, without further explanation, annulled the terms of the YLB-ACISA agreement. Although, under different circumstances, the cancellation of the contract with ACISA might have calmed the political situation in Potosí, at least temporally, under the actual circumstances, it made no impact on what had become a coup d'état, one in which some of the country's military leaders had thrown in their lot with opposition leaders such as Luis Camacho, the firebrand civic militant from Santa Cruz.

A week after the MAS government cancelled the YLB-ACISA agreement, the crisis reached a turning point with the release of an audit of the October 2019 election by the Organization of American States (OAS), which claimed to have found irregularities in the voting process. Although this audit was itself later discredited on a number of methodological and political grounds, its impact on the crisis was immediate. On the same day, General Williams Kaliman, the commander-in-chief of the Bolivian Armed Forces and a graduate of the US Army's notorious School of the Americas, called on Morales to resign. This major escalation was the first time in almost forty years that Bolivia's military had intervened in national politics, and it carried with it the very real threat of a military takeover of the country.

With echoes of the dark decades of military dictatorship suddenly ringing loudly, Morales announced his resignation on November 10 and went into exile to Mexico. The entire MAS government likewise collapsed, with some officials also fleeing the country and others taking refuge in the Mexican embassy in La Paz. Within days, leaders of the coup descended on La Paz and "retook" the *Palacio Quemado*, the historic seat of the Bolivian presidency. A particularly striking image showed a broadly grinning and self-satisfied Luis Camacho arriving at El Alto airport from Santa Cruz, surrounded by a phalanx of bodyguards and other members of his *cruceño* entourage. An infamous year of persecution, corruption, right-wing neoliberalism, and anti-Indigenous propaganda was unleashed.

Looking very much like swaggering interlopers who had stolen the keys to the halls of power, the large group of *golpistas* gathered for an inauguration ceremony, during which the figurehead "interim" president, Jeanine Áñez, an obscure evangelical Christian opposition senator from Beni Department, was sworn into office. She was followed by a chilling rogues' gallery of ministers over the following days and months, including the defense minister, Luis Fernando López, who would be implicated in massacres against crowds protesting the coup government; Branko Marinković, the exiled former leader of the *Comité pro Santa Cruz*, who had played a leading role a decade before in the Santa Cruz separatist movement, which might have included a plot to assassinate Morales; and the ghoulish Arturo Murillo, who became the face of the coup over the coming year, as he promised to "hunt down" Morales and other high-ranking MAS officials for "terrorism," "sedition," and other bogus charges drawn from the handbook of Bolivian dictatorship.

Before describing the various consequences of the coup and its aftermath for Bolivia's lithium project, something must be said about the possibility of a more sinister connection between the two—namely, that the Morales government was overthrown in a coup that was only nominally about political power and legitimacy and was actually about control over the country's lithium reserves. Morales, García Linera, and many others made the claim, then and later, that lithium was the real reason for the coup, and this is an idea that has the weight of history behind it.

Given that coups d'état throughout Latin American history almost always involve struggles to control resources (copper and telecommunications in Chile, vast tracts of agricultural land in Guatemala, bananas in Nicaragua—the list goes on), the case of lithium in Bolivia would seem to fulfill all the criteria for such an explanation. It is a resource whose global value was increasing rapidly; there was a more-or-less socialist government in power whose plans to industrialize lithium either excluded or put significant restrictions on the role of private capital; US companies had never been seriously considered during any of the international tenders, while Chinese companies had enjoyed something like privileged access; and, perhaps most importantly, almost all of Bolivia's twenty-one to thirty-one million tons of lithium, and thus hundreds of billions of dollars in potential earnings, remained in play.

And yet, as it turns out, sometimes there *is* smoke without fire. If the coup of 1980 in Bolivia, which brought the dreaded Luis García Meza to power, can be accurately described as the "cocaine coup,"[38] the same cannot be said for the role of another white powder during the coup of 2019–2020. While clear-eyed observers like Farthing and Becker rightly point to the broader and longer-term struggle in Bolivia between advocates of neoliberal privatization and advocates of "nationalization with dignity," as PA 25 puts it, my own research suggests that control over the country's lithium reserves in particular was not a factor—or, at least, not a significant one—behind the events of October and November 2019.

Perhaps surprisingly, my case against the idea of a lithium coup begins with an American spy. In late December 2019, in the midst of the unfolding social and political violence, I returned to Bolivia for the second extended period of ethnographic research during the four-year project (2019–2023). The routing during these years usually required a layover in Bogotá, Colombia, sometimes of several hours, sometimes overnight. After a phase of promising pilot research in August 2019, during which our team had received authorizations and research access to the lithium facilities from YLB, we had watched the coup of October and November from afar with deep apprehension as state institutions—including YLB—were purged of MAS loyalists, which necessarily meant the firing of many of the administrators, technicians, and support personnel who had been responsible for the day-to-day operations, both at YLB headquarters in La Paz and at Planta Llipi at the salar. I thus made the long journey from Lausanne, Switzerland, to La Paz filled with a profound sense of both sadness for the plight of the people and country I cared so much about and trepidation about the status of the lithium project. Was it about to be transformed beyond all recognition?

As I waited to board the final flight from Bogotá to La Paz, I happened to notice a man waiting at the same gate. My anthropological sensibility, which can lead me astray as often as it can point me toward something that I experience as revelation, narrowed on him immediately: forty-something, most likely American, traveling alone, traveling lightly, and perhaps most revealingly (and admittedly strange, or strange to admit), very well built for his age. He didn't look like a solo tourist, or a businessperson, or an American evangelical on a religious mission (they usually travel to Bolivia in groups, often wearing matching shirts), or even a fellow anthropologist (too much time pumping weights). My first thought was that he was a soldier of some kind. But he was traveling alone, and he was standing in the business class line.

So I sidled up to him and tried to make small talk. Did he speak English? Was he traveling to La Paz (although I knew the answer)? Was he an American? His demeanor as I posed these ice-breakers was almost painfully reticent; I tried to play the part of the amiable and much-too-talkative fellow American in order to draw him out. As he reluctantly answered my questions with clipped answers while giving me a steely look, I thought he might even just turn away (or worse). But I had the advantage that we were both in line to board the plane and he didn't really have anywhere else to go. In a stroke of ethnographic luck, the V-shaped stranger suddenly softened.

Was he visiting Bolivia on an organized tour (I also knew the answer to this question . . .)? *No, I work in La Paz.* Oh, really, do you own a business? *No, I work at the embassy.* Are you an FSO?[39] *No, I work in a different section.* Which one? (And then, the answer I had been expecting.) *I'm the defense attaché for the US embassy.* My anthropological sensibility started buzzing even louder. The defense attaché program is a key branch of the US intelligence ecosystem, part of the Defense Intelligence Agency (DIA). Along with agencies like the CIA and the NSA, the DIA participates in the full spectrum of US intelligence operations, including surveillance; monitoring; covert funding; and, especially in Latin America, the ignominious history of supporting regime change that would benefit US political and economic interests. Defense attachés are "overt" spies, meaning that their work as intelligence officers, based in US embassies around the world, is acknowledged, something that, among other things, allows them to claim diplomatic immunity. My ethnographic good fortune continued: the airplane wasn't ready for boarding and seemed to be delayed, perhaps for quite some time.

With more time on my hands, I began my soft interrogation of the American spy. I told him I was a researcher "studying Bolivia" (a useful anthropological gambit that is both true and usefully vague). Without using the word *coup*, I mentioned that I was very surprised that Morales had fled the country. I asked him if the US embassy was also surprised by the turn of events. *Yes*, he replied, *we had been following the situation closely, as you can imagine*. But didn't the embassy consider Morales to be hostile to US interests? (This was at the end of the first Trump Administration.) *In certain areas* [maybe drug policy?], *but we had no idea that the conflict over the election would turn in this direction*. I looked at him as if to say, "Sure, you are an American spy; that is precisely what you'd say if the US had used 'active measures' to support the anti-MAS opposition coalition"—active measure which might very well have included assurances to the coup leaders and Bolivian military of US assistance once the MAS government had been ousted.[40]

But his responses didn't seem rehearsed or evasive. The top "overt" US spy in Bolivia seemed genuinely perplexed by the recent turn of events and noticeably cool on the status of the coup leaders. For example, by late December 2019, the Áñez government had already announced to great fanfare that the government had reestablished ambassadorial relations with the US, which had been broken as far back as 2008. However, the defense attaché said this wasn't true; as he explained, the embassy was in no rush to embrace the "interim" Bolivian government, even if the coup leaders were eager to embrace the US.

And then I got to the question that I really wanted to ask. I said that among the topics I was studying, I was particularly interested in the country's lithium project. I strained to detect a knowing look, something that might betray his knowledge of the *actual* forces behind the coup—and the US's nefarious involvement in it. But there was nothing. He just nodded like he had never heard of lithium before in his life, or at least had never given the critical mineral much thought. Still, I couldn't resist: Was the political conflict really about lithium? Without hesitation: *No, not that we know of.*[41]

I realize that this is hardly dispositive evidence against the claim that control over Bolivia's lithium reserves was the real reason that the shadowy "21F" opposition coalition set in motion the series of events that led to the overthrow of the Morales government—but it is not negligible either, especially from an anthropological perspective which recognizes that data can span a very wide range indeed. Even if I fully agree with Farthing and Becker's general assessment that a desire to gain control over the country's resources—

and extensive hard currency reserves—was an important factor among at least some of the coup leaders, I don't believe that lithium was one of these targeted resources.

It must be remembered that by late 2019, many of the key elements of the lithium project had already been put into motion, including the inauguration of the industrial potassium chloride plant. The initial stages in the construction of the adjacent industrial lithium carbonate plant were also already underway. Moreover, although the Morales government's desperate move to cancel the YLB-ACISA agreement meant that the status of this major commercial project with the Germans was in doubt, this turn of events couldn't have been foreseen by putative lithium coup plotters hoping to take over the process.

In addition, there is no question that although the pro-US cabal of actual *golpistas* was perhaps ideologically more inclined to do business with US companies than with Chinese companies, the extensive investments by Chinese firms across a wide range of Bolivian industries—from construction to transportation to agriculture—had been very good for private business in places like Santa Cruz. Thus, the reality of MAS's long-term policy of selective nationalization coupled with support for private industry in the country—in sectors as diverse as mining, cement production, and agribusiness—means that the events of October–November 2019 should not be viewed as a simple struggle between capitalist and socialist political economic visions.

But even if the coup was not principally *about* lithium, it nevertheless had a profound impact on the project on a number of levels: economic, institutional, political, and technical. And even if the coup plotters were not initially driven by a desire to take over the lithium project, they certainly turned against it in different ways during their months in power. It must also be remembered that Bolivia was gripped by the tragedy of the Covid-19 pandemic for most of the period in which the "interim" government was in office, something that only magnified the depth of its brutal incompetence.

In January 2020, with the YLB-ACISA agreement in tatters and the country shaken by a pervasive sense of collective shock, regardless of ideological orientation, the first impacts of consequence for the lithium project were institutional. The Áñez government ordered a purge of the state lithium company along political lines. Given that YLB was a shining jewel in MAS's political economic crown and was thus staffed by hundreds of workers of different kinds who supported the political dimensions of the lithium project, this meant that unprecedented numbers were fired, regardless of expertise. Indeed, from the first days of the new year, YLB lost many of its critical employees, including engineers, project managers, and policy advisors, starting with the widely respected director of YLB, Juan Carlos Montenegro, who was replaced by Roberto Saavedra Villarroel, a previously unknown electrical engineer from Potosí with no discernible experience or expertise with lithium.

During these early weeks of January, I watched an incredible scene unfold at the headquarters of YLB, which is located on the top floors of the Edificio Hansa (1979), a towering skyscraper in central La Paz designed by one of Bolivia's most prominent architects, Juan Carlos Calderón (a 1957 graduate of Oklahoma State University of all places), whose other iconic buildings in La Paz include the so-called Communications Palace and the Plaza Hotel, among many others. Given the sweeping and sudden nature of the dismissals, YLB needed to replace a significant part of its entire workforce, from the administrative staff at the headquarters to drivers at the facilities at the salar.

As I left the elevator, I was confronted by a long and diverse line of job applicants, a line that snaked both up and down the internal staircase. There were at least fifty people at that moment waiting to make their way into YLB's main entrance, where they would present two copies of their materials, both of which are stamped with the date, which serves as proof that the materials were at least officially received. One man in his thirties was applying for a job in communications; a younger woman, probably in her late twenties, was applying for a secretarial position; another man, also in his thirties, was responding to an announcement for an open position at Planta Llipi in logistics. Along with the admirable stoicism that is required to wait in long lines, most people I spoke with knew very little about YLB, the lithium project, or what their particular job might entail. The spectacle of the mass sackings at YLB followed by the mass of job applicants spoke to the way in which the chaos and turmoil of the coup played out in the quotidian drudgery of institutional life.

But they played out in more spectacular ways as well. In a completely misguided effort to placate its allies in COMCIPO, especially its leader Marco Pumari, whose profile in the anti-MAS opposition was closely tied to that of Luis Camacho, the Áñez government decided to replace Roberto Saavedra as director of YLB with Juan Carlos Zuleta, even though Saavedra had been hired less than a month before. Unlike Saavedra, Zuleta *was* a well-recognized figure with long-standing expertise in lithium, but his background ensured that his tenure in office would be doomed from the beginning. An economist from Potosí with graduate training from the University of Minnesota and the New School for Social Research, Zuleta was the most prominent and vocal critic of YLB and the MAS government's entire program for lithium industrialization. Zuleta had worked for many years as an advisor to the private Chilean lithium industry and was, before the coup, also advising COMCIPO in its campaign against the YLB-ACISA agreement.

From the moment he took the reins of YLB in mid-January 2020, Zuleta's appointment provoked fierce resistance from a number of local movements in the salar region, including the Civic Committee of Uyuni and FRUTCAS, which viewed COMCIPO, the city of Potosí, and COMCIPO's lithium advisor Zuleta, as antagonists in the simmering struggle over future royalties and regional influence. And in the case of Zuleta, FRUTCAS claimed that the new director of YLB was actually still working on behalf of his Chilean clients to undermine Bolivia's state lithium project and open it to foreign capital. Just days after taking office, Zuleta attempted to visit the facilities at Planta Llipi, but he was blocked by dozens of FRUTCAS members, who declared a state of local emergency. As the federation put it, in announcing the general mobilization, "We repudiate and categorically reject the designation of Juan Carlos Zuleta as director of . . . YLB on the basis that he is a pro-Chilean agent and principal enemy of the 100% state industrialization of lithium [in Bolivia]."[42] Within three weeks, Zuleta himself was removed by the dangerously shambolic Áñez government.

Throughout the rest of 2020, the situation in Bolivia went from chaotic to catastrophic; despite its many abuses, the coup government was equally swept along by the wave of the public health crisis, which touched institutions and families irrespective of political dividing lines. But at ground zero for the lithium project, the YLB facilities at Planta Llipi and on the salar, the convergence of the coup with the Covid-19 pandemic had a devastating effect on everything from construction to the viability of the industrial process itself. One of the most well-placed observers of these repercussions was Guido Quezada Cortez, a chief geologist for YLB and the leader of the team that had conducted the blockbuster reanalysis of the salar's lithium reserves (see introduction). Like everyone else, regardless of seniority or technical indispensability to the state lithium project, Quezada had been sacked in the aftermath of the coup.

He described what happened at the salar as nothing less than "sabotage" by the Áñez government. From his perspective, the lithium project had become the leading symbol of MAS's economic vision, despite the actual difficulties over the years in moving forward toward the industrialization of lithium carbonate. Although the Morales government had invested almost $1 billion in the lithium project by 2020, the Áñez government decided to use YLB and the lithium facilities as a kind of negative showcase: if the project collapsed, it would support the argument for privatizing state industries across the board. As we have seen, although this plan was very likely not a central factor in the coup itself, it became part of an emerging strategy in the months *after* the coup leaders had taken over the institutions of government.

Apart from the purge of employees at all levels, which obviously brought YLB operations to a crawl, the Áñez government ensured that the functioning of the industrial process at the salar was severely curtailed, which had a particularly profound impact on the evaporation ponds. Quezada learned through some of his few former colleagues who remained at Planta Llipi that technicians had been ordered not to maintain or even monitor the evaporation process, in which brine is pumped from one pond to the next. Given that the evaporation process, as it proceeds from the first pond to the last, can take anywhere from twelve months to two years, depending mostly on weather conditions, this meant that the ponds were becoming increasingly degraded and at risk of being unusable—either to produce potassium chloride or lithium carbonate.[43]

As Juan Carlos Montenegro explained, the year after the coup had been a "lost year" in more ways than one. Montenegro, who had been the inaugural director of YLB and a key part of the lithium project right up to the moment he was fired by the Áñez government, said that the attempt to slowly destroy the evaporation ponds and the purging of YLB employees, among other acts of commission or omission, had a clear purpose: "They tried to make YLB disappear, to not be able to generate any income, to not be able to pay its debts to the [Bolivian] Central Bank, to go into default, and then to have to declare bankruptcy as a state company."

As it turned out, however, the ponds on the salar were not destroyed, although they were so degraded that YLB technicians would later need to spend almost a year to restore the evaporation process. And YLB did not go bankrupt, although, as will be seen in the next chapter, the chronic absence of a return on the state's investment and the chronic delays in developing the capacity to produce lithium carbonate at an industrial level would continue to trouble YLB—and increasingly shape lithium policymaking.

For the remainder of the *annus horribilis* of 2019–2020, the Áñez government proceeded to unravel in the midst of widespread political persecution, undisguised racism and contempt for the cultural legacy of the Morales years, and utter hostility toward the institutional framework of the plurinational state itself. Perhaps exhausted by its own sheer lunacy and ruthless ineptitude, the government allowed elections to take place in October 2020, during which a new MAS leadership was returned to power with the promise, among other things, to focus anew on the lithium project and return it to its place as the political economic cornerstone of a resurrected process of change.

CONCLUSION: "FIRST LITHIUM" AND THE PERILOUS ALLURE OF POTENTIALITY

In her ethnographic study of petroleum imaginaries and indeterminacy on the African island nation of São Tomé and Príncipe, Gisa Weszkalnys argues that the contours of resource futures in the present are marked by a "double obscurity."[44] On the one hand, the theoretical extent of latent resources is established through various technics of prospecting, exploration, and estimation, while the actual extent remains geologically obscure, hidden, and indeterminate. Even so, this condition of geological obscurity provides an opening for a wide range of decision-making and investments, from construction projects—both those related to resource extraction and those adjacent to it, like tourist infrastructure—to new directions in economic policy. But on the other hand, resource potentiality is framed by a second obscurity, one that conceals the future economic benefits of a resource that has yet to be extracted and converted into an economic asset.

As Weszkalnys shows, the implications of remaining perpetually suspended in a state of what the petroleum industry describes as "first oil"— meaning the uncertain period of transformation in anticipation of finally beginning production—include the need to impose constraints on national economic planning and the increased likelihood of social and political conflict. This is because governments that allow themselves to become trapped in conditions of first oil must continue to make specific promises about national economic and social development that become increasingly hollow as months give way to years, and, as in São Tomé and Príncipe, years give way to decades. And yet, the embrace of resource potentiality by states is often irresistible, despite the dangers of double obscurity and the risk of increased social and political disenchantment. These dangers and risks are measured against the *allure* of potentiality, the ways in which its very indeterminacy can itself become a resource with political and ideological value.

As we have seen in this chapter, the contested unfolding of the lithium project in Bolivia has likewise been infused with the perilous allure of potentiality, a prevailing motif that helps conceptualize and bring much-needed nuance to our understanding of what is in fact a *trajectory* of contestation. Across the entire sweep of Bolivia's "post-neoliberal" history, from the coming to power of MAS in 2006 to its return to power in 2020, different administrations and state institutions have grappled with the unsettling indeterminacy of "first lithium." But if the condition of first lithium is characterized by its own double obscurities—both geological and economic—it must also be understood through the dual nature of its potentiality. Vast amounts of both state resources and geopolitical capital have been expended based on the promise that Bolivia would be able to industrialize its immense lithium reserves and, in the process, break the chains of its "colonial and republican heritage."

But as this chapter has also revealed, conditions of first lithium in Bolivia have facilitated social and political conflict as much as they have imposed constraints on a national economy organized around the logic of "productive sovereignty." Although COMCIPO's general mobilization against the YLB-ACISA agreement preceded the coup of October–November 2019, the fact that the agreement was announced a full decade after ground was broken at the salar for the lithium carbonate pilot plant only fueled the ability of COMCIPO militants to cast doubt on the agreement, to describe it as yet another false promise. In this way, the dangerous indeterminacy of first lithium sparked a local and regional conflict that converged with wider opposition forces, giving them yet one more rhetorical weapon to be wielded when the moment of usurpation arrived.

And yet, in thinking about how the social, political, and economic conditions of resource potentiality come to mirror the indeterminacies that envelop the resources themselves, it would seem to matter *which* resource is at issue and *how* the particular resource fits within wider policy and ideological formations. In this sense, the politics of first lithium in Bolivia—however perilous, as we have seen—are still tightly intertwined with the global energy transition. To make promises about the transformative potential of lithium industrialization is to make promises about a future that is meant to be part of the solution to the climate crisis, however unlikely that solution may be in practice. This fact blunts the sharp edges of disenchantment that make first lithium, like first oil, such a compelling social problem. It also means that the Bolivian government will continue to put the lithium project at the very center of its own plans for the future, despite its troubled history.

─────────

THREE

Flexible Extractivism and Historical Reckoning

Continuing southward, in a town called Jayu Cota, [Tunupa] dug up the earth again, this time to leave her milk for her youngest son, who was following her. This place is now a salt flat of reddish color. . . . She continued her journey until arriving at the Uyuni salt lake. . . . In this region, she met two handsome young men—Cora Cora and Achacollo [Big Mountain]. . . . [T]he two young men began fighting each other for Tunupa, and a war broke out. With a large catapult, Cora Cora wounded Achacollo's heart, causing great bleeding. For this reason, today that mountain appears completely dry. But Achacollo also struck a blow to Cora Cora, wounding him in the bladder and opening many holes. This mountain today has several streams coming from its interior. Thus, both of them died for Tunupa's love, and from that point on, Tunupa stayed in the region.

From *"The Myth of Tunupa," recorded from oral*
tradition by the anthropologist ramiro molina rivero[1]

STANDING AT THE CENTER of the industrial complex on the Salar de Uyuni—which includes processing plants, storage warehouses, an assortment of maritime shipping containers (several owned by the Danish company Maersk), offices for YLB employees, dormitories for construction workers, and transport pipelines that extend to the distant evaporation ponds like tentacles—one also can't help notice the most prominent landmark on the horizon, a mountain peak whose bottom half disappears below the curvature of the Earth. This is the mighty stratovolcano Tunupa, which looms over the entire salar region at around 17,500 feet. Based on the surprisingly vague criteria of volcanology, Tunapa is *probably* extinct rather than simply dormant, given that it was last active around 1.5 million years ago.

Culturally, the importance of Tunupa testifies to the region's ritually dense landscape, the fact that topographical features of all kinds—volcanoes, ravines, rocky outcrops, hot springs, hills, the salares themselves—have been infused with meaning by the Indigenous communities of the vast surrounding altiplano for over a thousand years. But the iconic presence of Tunupa, the goddess whose breast milk created the Salar de Uyuni, also testifies to something else: the fact that Bolivia's lithium project is located in one of the most geologically active zones in the world, where the Ring of Fire's Nazca plate continues to thrust itself from the west under the South American continent, a process of tectonic confrontation and terrestrial violence that forms the Andes and altiplano and, at the same time, has been responsible for some of the largest earthquakes in human history.

The Chinese companies who were contracted to build the industrial lithium carbonate plant for YLB, working with a design submitted by the German process engineering firm K-UTEC, were therefore confronted with a serious problem: how to build a facility with the capacity to produce fifteen thousand tons of lithium per year that both was capable of withstanding a major seismic event and was kept within the relatively modest $96 million construction budget. In the months before the 2019 coup d'état in Bolivia, construction on the plant moved at a snail's pace in large part because of the need to install earthquake-tamping technology in the different buildings.

One YLB technician explained to me that layers of latticed steel rebar had to be built under the industrial structures, creating a pliant base that would at least theoretically absorb energy within calculated parameters in the event that the plant was rocked by the kind of earthquake that had earned the Ring of Fire its historic reputation for meting out tectonic peril. Yet by 2021, well after the new MAS government had retaken the reigns of the lithium project, the need to ensure that the new industrial lithium plant was at least earthquake *tolerant* continued to create major delays in the construction timeline—delays that went well beyond anything related to the condition of the evaporation ponds, the problem of rebuilding a decimated YLB workforce, and the turmoil created by a conveyer belt of changing leadership at YLB headquarters in La Paz.

Whether or not the malleable architecture of Bolivia's industrial lithium facilities will ever be put to the test is something that remains to be seen. But I begin this chapter by describing this problem as a way of shining a critical light on a wider dynamic, one that continues to shape the lithium project in ways both big and splashy and also nuanced and largely unacknowledged. The return of the MAS government in late 2020 under the leadership of Luis Arce coincided with a deepening realization at all levels, from the engineering staff at the salar to financial managers worrying about YLB's debt in the

FIGURE 9. Construction underway at the Salar de Uyuni, 2020. Photo by author.

Bolivian Central Bank, that the actual pursuit of lithium industrialization confronted *resistance* at many levels—not only at the level of rhetoric—only some of which had been taken into consideration at earlier stages in the process. This realization eventually had a recursive effect on the lithium project itself; as will be seen, it led to both more obvious changes in industrial strategy and to a more subtle transformation to the project's underlying political economic framework.

The ethnographic account of these important shifts, which carry lessons for understanding decisions taken by other states caught up in other extractivist webs of the global energy transition, reveals that energy and economic policymakers in Bolivia responded to a spectrum of resistance that included the peculiarities of brine composition in Bolivia; shifting weather patterns in the salar region; the elusive pursuit of new technologies for lithium production; and a profound tension between the different imperatives of resource nationalism and the limits of national industrial and scientific capacity. In this way, the manifold exigences of the lithium project in Bolivia—which were never static over time—came to exert their own collective agency, a force that transcended the shifting political fortunes of the different national governments in Bolivia.

Moreover, the effects of this agency were felt in both directions: material, technological, and political forms of resistance compelled shifts in political economic strategy and these shifts then, in turn, shaped how both existing and new expressions of resistance within the lithium project were addressed at remarkably delimited scales. As I argued in the introduction, these shifts gave rise to a distinct framing for the lithium project, one that can be understood as the next or, perhaps, final stage of neo-extractivism: *flexible extractivism*. This describes the ways in which the Bolivian government and its state lithium company turned away from the grand ideological ambitions of the lithium project's early years to embrace an ethos of production at all costs and a willingness to acknowledge the difficulty in reconciling competing demands for environmental protection, regional and local decision-making, and national economic growth.

This chapter unpacks the material, technological, historical, and political dimensions of flexible extractivism, the ways in which different forms of reckoning with and through the lithium project led to transformations in the country's relation to both its own economic future and to wider energy and climate policymaking. In order to understand how this reckoning unfolded, the chapter begins at the most important sites of geological struggle and productive anxiety: the YLB facilities at the Salar de Uyuni, where quite granular problems of labor, technological capacity, and climate, among others, were eventually taken up and reframed as problems of political economy that demanded a new approach at the level of national energy and industrial policy. A first section examines both the geological and phenomenological complexities of the Salar de Uyuni, where YLB technicians struggle to transform its fractal mysteries into a resource of global importance. A second section narrows in on one particular dimension of this complexity: the surprising technical and metaphorical significance of core samples, which are used to make projections about the unknowable whole from its somewhat-more-knowable parts.

The chapter then moves out to show how the impact of these geological, technological, and historical difficulties led the MAS government of Luis Arce to revise its national economic blueprint, while attempting at the same time to preserve at least a symbolic adherence to the principles of Patriot Agenda 2025. As will be seen, almost immediately after regaining political power in October 2020, the government set to work on a new economic and social development plan that sought to reconcile the tension between existing ideological principles and the more pressing demand to generate new sources of state revenue, especially from the stalled lithium project.

This attempt at a balancing act took shape through the 2021 "Plan for Social and Economic Development," which became the political instrument for the institutionalization of flexible extractivism in Bolivia. Borrowing a page from the country's well-established approach to gas and oil development, a move that a number of interlocutors described as the "hydrocarbonization" of the lithium project, the plan deepens the focus on production, reaffirms the need to pursue joint partnerships with foreign companies, and refuses to abandon resource extraction itself, despite criticism from intellectuals, environmental activists, and some local communities.

The chapter concludes by reflecting on what the rise of flexible extractivism—within the Bolivian lithium project and more generally—has to say about the future of the ongoing global energy transition and its tenuous connections with shifting or ambiguous notions of climate justice. If flexible extractivism is justified not only as a pragmatic rethinking of political economy, but, even more, as an expression of climate justice from the South, what does this bode for the decarbonization mandate at the heart of international climate policy? Can resource extraction *ever* be understood as both a form of resistance to what Loreta Tellería has described as the "green hegemony" and as a fundamental plank in the global energy transition at the same time?[2]

THE EYES OF THE WHITE DRAGON

In 1873, in a remarkable coincidence, two young researcher-explorers arrived in Bolivia, both of whom would go on to play important roles in the history and politics of the Salar de Uyuni. Even more coincidental, both were almost the same age. Manuel Vicente Ballivián was the twenty-five-year-old son of a family of Bolivian exiles who had spent many years in Peru, Spain, England, and France. His father was the writer and historian Vicente Ballivián y Roxas, who had labored in various foreign archives in an effort to produce a complete history of Bolivia's pre-Columbian and colonial periods. Like his father, Manuel Ballivián had studied at universities in England and France and returned to Bolivia as a fluent speaker of both English and French (in addition to his native Spanish) with ambitions to help develop the scientific profile of the young republic, which was at the time less than 50 years old.

Jack "John" B. Minchin was twenty-four years old when he arrived in Bolivia. Minchin—who would come to be known as "Juan" Minchin—had traveled along an equally fascinating path. Minchin had been born on a country estate in County Tipperary, Ireland, before emigrating to New Zealand in 1852 with his family at the age of three. When he was sixteen, he was sent to London for his university studies. After completing a degree at King's College, London, he worked as an engineering apprentice and was certified as a civil engineer in 1870. During his apprenticeship, he gained valuable experience working on the Tower Subway (a pedestrian passage under the Thames River that closed in 1898) and on different railway projects. In 1873, the young Irish engineer applied to work for the British government, which sent him to Bolivia to conduct engineering and survey work in collaboration with Bolivian colleagues.[3]

Ballivián would go on to become one of the most influential Bolivian academics and scientific administrators of the late nineteenth and early twentieth centuries. Besides overseeing a number of historic national censuses and later serving as Bolivia's agriculture minister, he coauthored (with Eduardo Idiáquez) the magisterial *Diccionario Geográfico de la República de Bolivia* (1890) and became the director of the Geographical Society of La Paz in 1897. During his long career, Ballivián participated in numerous regional and international scientific congresses and was perhaps the most cosmopolitan Bolivian scientist over these decades (he died in 1921).

Minchin's future trajectory was also luminous in its own way. He did perform engineering work on behalf of various Bolivian agencies, including a study of potential railway networks that took him to the salar region. However, by 1880, he had left government work to focus his expertise on mining. By the 1890s, Minchin had become wealthy as the owner of mines in Oruro and Uncía (which is in the north of Potosí Department). After suffering a stroke, Minchin left Bolivia for good in 1911 and died in 1922 in London. Before leaving Bolivia, he sold all of his mining stakes for the enormous sum of £600,000 (about $75 million today).

But before the remarkable lives of Ballivián and Minchin could take their full courses, they intersected: in the mid-1870s, soon after they had both arrived in Bolivia, the two traveled together from Lake Titicaca to the Salar de Uyuni, where they conducted a geological survey of a number of lakes and salares. Although the two young surveyors didn't reach any definitive conclusions, they were the first to suggest that the altiplano region between Lake Titicaca and the Salar de Uyuni consisted of a vast basin of interconnected lakes and evaporated lakes that had formed over tens of thousands of years. After this early survey, the two went their separate ways. Then, between 1907 and 1913, a geographer from Yale University named Isaiah Bowman returned to the question of what he called the "lake system of the Bolivian Plateau."[4]

Beginning with research conducted for his doctoral thesis, Bowman eventually proposed the existence of a number of giant lakes that dated from "glacial times" (radiocarbon dating wouldn't be invented until the 1940s). In his 1914 paper for the American Geographical Society, Bowman names the largest of these prehistoric lakes "Lake Minchin," "after the late Juan B. Minchin of Oruro, whose studies of forty years ago mark the beginnings of our scientific knowledge of the Bolivian basin."[5] Bowman names a second of the giant prehistoric lakes "Lake Ballivian [sic]," "in honor of Don Manuel Vicente Ballivian, Bolivia's most distinguished scholar."[6] Later studies would confirm and refine Bowman's model for understanding the geology of the salar region.[7] Lake Minchin slowly evaporated between about forty thousand and thirteen thousand years ago, dividing into a number of smaller lakes, some of which themselves evaporated, including the one that became the Salar de Uyuni.[8]

Today, the Salar de Uyuni has a surface area of roughly 10,500 square kilometers, making it by far the largest evaporated lake ("salar" in Spanish), or salt pan (also sometimes called a "salt flat"), in the world.[9] If the area of the Salar de Uyuni contained an entire country, it would be larger than Lebanon and just smaller than the island of Jamaica. It is also the largest nearly flat surface on Earth by a considerable margin. This fact brought researchers *again* from the US Geological Survey to Bolivia (in 2002) as part of an effort to measure the coordinates of the salar's surface. They did this by driving 4×4 vehicles equipped with GPS antennae back and forth across the salar, which gave them precision kinematic data for the entire area.[10] The reason the USGS wanted precise measurements of the salar's surface was so that it could be used as the primary location on Earth for calibrating the altimeters of orbiting satellites, including NASA's ICESat and ICESat-2 and the European Space Agency's CryoSat-2, all of which are (or were, in the case of ICESat) used to study the effects of climate change on ice sheets in the Arctic Sea, Antarctica, and Greenland.

Although the salar was shown to vary in elevation by less than one meter across its entire extent, the 2002 USGS study revealed another extraordinary feature: it detected small but patterned undulations of only forty centimeters at the exact places where the Earth's gravitational field passes over mountain ranges buried several kilometers below. This is the same phenomenon—called the "geoid"—that occurs in the world's oceans, which bulge at places on the surface where the gravitational field passes over deep underwater seamounts.[11]

But what about the geology of the salar itself and, more important, the lithium-rich brine that flows underneath its surface? The thin upper layer of the salar, extending to the surface, consists of halite, or rock salt. Because of wind coupled with the geothermal effect of brine being pushed to the surface through fissures and then evaporating, the salar is covered with tiny halite crystals and its famous desiccation polygons, which lend vast tracts of the salar the appearance of a kind of geometric intentionality.

Based on landmark surveys conducted by Bolivian geologists in the mid-2010s, which, among other things, led to the blockbuster revision to the estimates of the country's lithium reserves (see the introduction), scientists have been able to model the salar's lithostratigraphy to a depth of fifty meters (see below). These complex strata are coursed by brine, a highly saline fluid that first enters the salar as freshwater from a variety of sources, principally from the southeast, where the Río Grande delta meets the edge of the great evaporated lake. In addition, the salar region receives about two hundred millimeters of precipitation a year during the rainy season, which lasts from November to March.

The accepted explanation for why the salar's brine is so rich in lithium remains the one proposed by a US-French-Bolivian research team in the early 1980s: that over the span of geological time, the region's intense volcanism has created unusually large deposits of rocks called ignimbrites, which contain volcanic glass. As a report explains, "Lithium becomes concentrated in the late phases of an igneous melt due to fractional crystallization . . . and volcanic glass . . . provide[s] an especially good source of lithium because lithium is so readily leached from glass."[12]

In the case of the Salar de Uyuni (and the other smaller Bolivian salares), the widespread volcanism of the Ring of Fire just happens to coincide with the complex hydrogeology of massive evaporated lakes, which nevertheless continue to absorb fresh water beneath their surfaces. This water is then slowly transformed into a mineral-rich liquid that has passed over the many different underground strata but is protected from evaporation by the salar's halite crust. Because of this, the liquid is able to flow through the salar's hidden layers for long periods of time, leaching minerals as it makes its unseen way: sodium, magnesium, potassium, boron, and lithium. This explanation for brine formation also applies to the salares of Chile and Argentina.

But even if the highly specific geology of the Lithium Triangle is similar at the level of brine formation, there are also a number of critical differences that have made Bolivia's lithium project a source of torment for YLB's frontline technicians. As the Bolivian geologists Guido Quezada and Nelson Carvajal explain, drawing on an earlier model introduced by a group of economic geologists salares can be divided into two categories: "mature" salares, which are "halite dominant"; and "immature" salares, which are "clastic dominant."[13] *Clastic* means sediment that consists of fragments, bits and pieces—that is, geological detritus.

Besides the relative difference in the quantity of halite versus clastic sediment, these two categories of salares are distinguished by levels of precipitation, localized tectonic activity, altitude, and the specific geological structure of the salar itself (which is technically a "closed basin aquifer"). In general, mature salares are found at lower elevations; receive much less precipitation; and are, relative to the second category, much less porous through their different strata.

By these criteria, the Salar de Uyuni is an "immature salar": it is located at 3,653 meters; as we have seen, it receives about two hundred millimeters of rain per year; and its different strata (which run at least as deep as fifty meters, but likely deeper) are marked by "high levels" of permeability and porosity.[14] As a point of comparison, a classic mature salar is the Salar de Atacama, the center of Chile's lithium industry. It is located at 2,300 meters, receives only twenty-five millimeters of rain per year (less than an inch), and is much less stratigraphically complex.[15] Beyond the ways in which the annual rainfall continually transforms the Salar de Uyuni's underground geology through mixing, the formation of fissures, and sediment "transportation," it also has profound implications for the process of industrial evaporation: Unlike at the Salar de Atacama, Bolivia's evaporation ponds are covered by clouds during the five months of the rainy season, even if the greater elevation means they receive more intense radiation when the sun *is* shining.

All of these hydrogeological complexities are confronted and experienced in particular ways by those charged with advancing the process through which the brine of the Salar de Uyuni with the highest lithium concentrations must be located, accessed with industrial wells, pumped from varying distances into ponds, and then carefully monitored as the brine becomes more and more concentrated through evaporation, which itself can take over a year under the best of circumstances. The challenges do not end there. The resulting salt compounds then must be purified through a process that becomes increasingly chemically intensive as it leads to the production of—ideally, "battery grade"—lithium carbonate.

However, as I would come to learn, from the perspective of both technological difficulty and what might be thought of as the phenomenology of lithium industrialization, the first stages in the process are much more harrowing than the final steps. In other words, grappling with the confounding, unperceived, rhizomatic structure of the salar is significantly more arduous than using an electron microscope to confirm that the crystals inside of a given unit of purified salt compound are lithium and not, for example, magnesium.

I experienced the power of the salar's materiality during a period of unforgettable ethnographic research with one of Planta Llipi's head geologists and his team of technicians and drivers. Beyond the varied technical demonstrations, he always had a more general objective in guiding me through a series of participant observations, which took place at different places across the zones of production on the salar and within YLB warehouses and facilities. More than anything, the geologist wanted me to understand the ways in which the work of lithium industrialization was much more than simply another state job. Rather, it was an epic struggle—one whose outcome remained uncertain.

When I was fortunate enough to have the chance to see this struggle through his eyes, even partially, the geologist was already a weathered, world-weary, and resilient old hand with decades of experience behind him working for the state mining company, COMIBOL. In his mid-sixties when I met him, he had moved to the state lithium project when it was still under the umbrella of COMIBOL. He had spent his career working in both state-controlled and private mines, across the full range of national governments, beginning in the final years of the Banzer dictatorship. From both an institutional and political perspective, he had seen it all. His long experience in the mining sector had provided him with decades of practical experience in all aspects of the geology of underground hard-rock mining, which has long served as the iconic category for grueling and even deadly resource extraction in Bolivia.[16]

But all of the geologist's expertise in the mining industry did not prepare him for what he faced when he was called up for duty in the great struggle with the salar. Like a battle-scarred veteran, he wanted to show me the front-lines in the great struggle, to give me a taste of the geological dangers—as a lesson, and also, perhaps, as a warning. To begin, we left the plant in an YLB

4×4, traveling along the raised asphalt road that connects the industrial facilities on the salar with the oldest YLB buildings at the salar's edge (where roads then lead to places like Río Grande, Uyuni, and the port cities on the Chilean coast).

At a certain point, and without any warning, the YLB driver veered off the road and onto the salar itself. With questions swirling in my head about driving without any landmarks or the possibility of navigating in relation to a hallucinatory horizon, the driver nevertheless accelerated—40 kilometers per hour (km/hr), then 60, then 80, then 100. The only way I could tell that we were suddenly flying across the salar's surface was by shooting a quick glance at the speedometer. Looking outside the window, from my seat in the second row, it seemed like we were almost standing still, despite the roaring engine.

Very quickly, however, it became clear that our driver knew exactly what he was doing—and where he was going. The geologist pointed to a number of faint objects in the distance, which did seem to get bigger as we got closer, something I found oddly comforting. (We apparently *hadn't* passed through a wormhole into another dimension.) Every so often, the driver made a series of sweeping arcs instead of driving in a straight line. These were not really turns as one would imagine them but rather long leans or veers—at 100 km/h. As the driver later told me, he drove like that in order to avoid areas of the salar's crust that were particularly soft. How he knew where these areas were in a sea of undifferentiated whiteness, he didn't say.

I tried to engage the geologist in conversation, beginning with a series of broader questions about the country's lithium project, questions that didn't seem to be of much interest to him. Perhaps they were too abstract or too politically sensitive? As I had learned during research with officials at YLB headquarters in La Paz, the political pressure on the state lithium company at the time was even greater than usual. Maybe this pressure was being felt at the local level, the level of material struggle, in ways that were making the frontline workers nervous—about what, exactly, I couldn't say, perhaps their own futures? But as it turned out, the geologist wasn't being cagey; he was thinking deeply about the salar, and in profoundly existential, Melvillian terms.

After brushing aside my different questions about national policy and the relationship between lithium and green energy, he gestured in front of him: *This is all I can think about . . . this is our white dragon.* Even though I assumed he had either intentionally or unintentionally substituted a dragon for the white whale in invoking the literary masterpiece, the meaning was clear

FIGURE 10. Exploring the enigmas of the Salar with an YLB geologist, 2019. Photo by author.

enough: like Ahab in *Moby Dick*, he too was in the throes of a contest with a powerful and ultimately inscrutable adversary, a contest that might very well be his undoing.[17]

After about one hour of driving, we reached our intended destination. The geologist got out of the 4×4 and told me to follow him closely, meaning *right behind him*. I soon understood why. I could see in front of us what looked like thermal pools, but in the meantime, we started to pass in between hundreds of smaller holes in the salar, which bubbled with heated brine. *You need to be very careful here*, the geologist said over his shoulder, *these* ojos ("eyes") *can be dangerous. You don't want to fall through.* We picked our way through the treacherous smaller ojos in order to reach what he really wanted to show me: a series of larger brine pools.

The three of us stood at what I hoped was a safe distance from the edge of the pools, looking into them without talking. I realized that I wasn't actually sure what the lesson for me was supposed to be, apart from the obvious spectacle of the salar's geological dynamism. The YLB geologist knew that I was conducting research on different aspects of the state's lithium project, so perhaps these thermal pools played a particular role? I then asked a dumb question: Did these thermal ojos hold more lithium or were they used somehow in the process of brine evaporation? Even as I was asking it, I knew how inane the question must have seemed: We had just driven far across the salar and there was not a pump or pipeline or any other evidence of industrial activity in sight. The geologist just shook his head and continued staring into the thermal brine pool. He had brought me to what he considered a paradigmatic site for reinforcing a fundamental message, one that he believed I needed to receive in order to really understand the lithium project. But it was a message that still escaped me.

That's when the geologist decided I needed to experience the message in a different way. He instructed me to follow him again, to stay closely behind as we walked back toward the 4×4. But then he suddenly stopped and got down on his knees. I stood behind him watching as he began poking his index finger into one after another of the smallest of ojos. Satisfied that he had found what he was looking for, he told me to get down on my knees, right next to him. *Put your finger into that ojo*, he told me, *then into that one*, pointing to two small holes about thirty centimeters apart, which were otherwise completely (to my eyes) indistinguishable.

I eagerly complied, now burning with intense ethnographic curiosity. I inserted my own index finger into the first ojo and almost immediately recoiled: the brine was not exactly scalding, but it was very hot. I laughed nervously and looked over at the geologist: Wow, that's really hot, I said, not really knowing what else to say. I then stuck my finger in the second ojo and recoiled again, but for a very different reason: the brine was icy cold, which was shocking in its own way. The message the geologist wanted to impart started to wash over me, and I stood up and waited for him to give it proper form.

You see, he began, *two ojos right next to each other. How can the temperature of the brine be so different?* He answered his own rhetorical question: *Because they come to the surface from very different places, from different levels that aren't connected at all. We don't have a map* [of these underground pathways]. *The brine is different everywhere.* I wanted to ask a number of broader questions in that moment. I wanted to know whether he would trace this microgeological social fact to wider problems at the center of the country's lithium project. But I didn't say anything more—the demonstration itself would have to suffice for now. As we retraced our path back to the YLB 4×4, the geologist didn't have to remind me where to walk. Almost like I was playing a game, I trailed him closely, walking in his footsteps.

The return trip was as before, just in reverse. We flew back across the salar, through the same high-speed sweeping arcs, except this time I could make out the vague outlines of industrial facilities in the distance and, eventually, the raised asphalt road. Looking out the window, I also noticed for the first time that we were making a track by driving over the omnipresent desiccation polygons; I wondered how long it would take for the seeping brine and altiplano winds to form them anew. But as we zoomed back in a kind of professional silence—the geologist having apparently discharged his duty to the visiting gringo anthropologist—I wanted to know more; I wanted, as the French phrase has it, to *boucler la boucle*, to close the loop.

From the back seat, I leaned forward and looked at the geologist. I couldn't think of a properly technical question, something to do with the salar's geology or with the process of brine extraction. So instead of a question, I thanked him for showing me the ojos. I think I understand, I said, how difficult it must be to work under these conditions. I waited, hoping he would reflect again on the nature of the epic battle he and his colleagues were waging against the white dragon. As the seconds turned into minutes, and the professional silence was restored, it became clear that the loop would remain open, which was, I realized later, entirely consistent with the lesson I had supposedly learned on the salar. There are no closed loops within the fractal complexities of the salar, just as there are no closed loops in the lithium project itself.

SUBTERRANEAN WORLDS OF INFINITE COMPLEXITY

Scholars of traditional hard-rock mining in Bolivia have developed a number of illuminating theoretical approaches for understanding how spaces of labor can become what Andrea Marston has described as sites of grafting: extractive topographies to which political ideologies, legal rights, and anti-Indigenous racism, among other categories, are affixed.[18] Marston draws particular attention to the subterranean topographies of mining, especially those in which the "strata of the state" are also levels of contestation between different visions for resource extraction. These include the tension between MAS's resource nationalism and its stated—and constitutional—commitment to Indigenous autonomy, which should ostensibly guarantee greater support for communal and cooperative mining.[19] Decades earlier, June Nash also explored the signal importance of the subterranean in Bolivia as a space in which both political economic grafting and cultural resistance take place.[20] As her research revealed, the tin mines of Oruro are underground realms of danger and uncertainty in which miners confront wider, even global, forces of capitalist dependency through everyday forms of cultural praxis and perseverance.

The subterranean likewise takes on wider importance for the extractive materialities of lithium industrialization, but in ways and with implications that are radically different.[21] Instead of tensions between contrasting political economic visions, the revealed strata of the salar shape quite another tension. The more precisely the YLB geologists penetrate the subterranean strata of the salar, the richer is the geological data that is found and the more difficult the industrialization of the lithium becomes. As I learned during research on other pieces of the geological puzzle at the salar, the process of core sampling might have played a key role in the dramatic upward revision to estimates of Bolivia's lithium reserves, but it also fed into a broader narrative about the salar's many formidabilities, a narrative that itself influenced national lithium policymaking.

As with the salar's other disorienting materialities, the stratigraphic also had to be understood through what became yet another unplanned exercise in sensorial ethnography. Among other things, my own encounters with the extracted salar reinforce the importance of perceiving materials—geological or otherwise—through their undeniable "liveliness" rather than as inert objects awaiting classification.[22] The liveliness of the salar became vivid once more during a series of observations within a frigid and dusty YLB warehouse, one of the original buildings constructed between 2008 and 2010, in the early years of the lithium project's launch during the first Morales government.

As we have seen, a team of Bolivian geologists conducted a five-year study of the salar between 2013 and 2018. Apart from the global significance of the study for the question of Bolivia's likely lithium reserves, this survey also generated extensive stratigraphic data based on both the surface areas sampled and, even more consequentially, on the depth of the core samples. For example, in 2015, the Bolivian team took a single core sample from a location about 38 km to the west of the village of Colchani, which has the most well-developed artisanal salt mining tradition in the region. This borehole reached an extraordinary depth of 460m using a precision diamond drill attached to a rig. The process, which had an equally historic "core recovery" rate of almost 100 percent, took two months to complete.[23]

This deep borehole yielded unique data about the geology of the salar, including its paleogeology—since material from 460 meters was dated to over three million years ago—and was a landmark accomplishment in the history of Bolivian science. Among other things, the Bolivian state geologists were able to say that "Survey E-036" was over twice as deep as a 1999 perforation (223 meters) made by a research team from Duke University, and almost four times deeper than a borehole dug in 1989 by the French team led by François Risacher during the same survey that had established 8.9 million tons as the baseline for Bolivia's lithium reserves (see chapter 1).

But as Guido Quezada explained when I interviewed him about a year after he had been fired by the coup government, the work of Survey E-036, as important as it was from a scientific perspective, paled in comparison to the work that took place both before and after. This multiyear phase in the study began with the design of a series of grids across more than ten thousand square kilometers of the salar's surface, grids that were then used to guide the placement of sample perforations of up to fifty meters. Although the team dug boreholes—using the same non-percussive diamond drill technology—across much of the salar, it focused its attention on the area around the Río Grande delta, where the subsurface brine has the highest concentrations of lithium. Quezada emphasized that it was not just the total number of boreholes that mattered but the total number of meters of sample material collected, which is a function of the number of holes times their average depth. In the region of the Río Grande delta alone, the Bolivian geologists managed to collect 3,595 meters of sample material from an average depth of fifty meters.

It was a very small portion of *this* material that I studied in the cavernous YLB warehouse, the centerpiece of a demonstration that was meant to—yet again—underscore the scope of the salar's intricacies and their consequences for the wider lithium project. Several younger YLB technicians had pushed together, end to end, a number of large wooden worktables, creating a single long surface. When I arrived, accompanied by an assistant geologist, they had already filled this surface with dozens of smaller segments from just one of the many fifty-meter core samples. The most obvious characteristic, at least to my non-geological eye, was the spectrum of multilayered colors, which marked the different levels of halite, clay, hardened silt, and gypsum.

But then the assistant geologist told me to have a closer look, to actually pick up one of the segments. When I did, I immediately understood what made it a critical piece of evidence in a way that the palette of colors did not. I raised one of the cylindrical sections up and felt its importance before I saw it. My fingertips disappeared into small holes and fissures everywhere I held it. I looked closer and saw that the core sample segment was riddled with these little tunnels, which interlaced the cylindrical sample without any apparent regularity—in direction, in depth, or in size. Some of the holes passed all the way through; some were more like deep indentations. The randomness remined me of the holes created by trapped carbon dioxide during the fermentation of Emmentaler cheese.[24]

But then it struck me: I was holding a tiny segment of infinite complexity in my hands. This one small cylinder from just one fifty-meters core sample was impossible to comprehend, not in itself, which was difficult enough, but as an infinitesimal piece of an illimitable whole. Any attempt at real apprehension vanished as soon as I tried to reinsert it in my mind's eye back to where it came from somewhere in the southern part of the salar, then magnify it by the salar's entire geological volume. *This* was the true measure of Bolivia's lithium materiality, at least at the point of brine extraction, since the mineral-rich fluid will always course through a subterranean world whose actual forms must remain ultimately unknowable, despite every effort at sampling, projection, and approximation. And it was in part the growing realization of this and other inescapable material realities that eventually led to the emergence of a new political economic orientation to the lithium project in Bolivia, and a greater willingness to come to terms with its fraught history.

THE HYDROCARBONIZATION OF LITHIUM

In the months leading up to the October 2020 general elections in Bolivia, a sense of inevitability started to creep into national political discourse. The combined effects of both a year of political trauma and the public and social tragedy of the Covid-19 pandemic had produced a pervasive weariness and a profound desire for change. The "interim president" Jeanine Áñez withdrew her anemic candidacy in late-September in an effort to shore up the other anti-MAS candidates. These included the odd-couple right-wing ticket of Luis Camacho, one of the leaders of the coup, and Marco Pumari, the head of Potosí's COMCIPO, who had gone on a hunger strike the year before to protest the YLB-ACISA deal (see Chapter Two). Former president and vice-president Carlos Mesa returned as a candidate with a softer anti-MAS orientation; he represented an option for a segment of the population that rejected the violence and shock of the coup but still wanted a government that was "ABM"—anything but MAS.

The problem with this strategy was that many Bolivians had already moved on from the controversy around the October 2019 elections and the question of whether or not Evo Morales had, in fact, won the election outright, despite what came to be seen as deeply biased charges of electoral fraud, including by the Organization of American States.[25] At the same time, compared with the complete and utter disaster of the year that followed MAS's ouster—the rank and barely disguised corruption at the highest levels; the bungling of the public health response to the pandemic; the attempts to erase all traces of pro-Indigenous policy; the mobilization of the courts and police in one of the starkest cases of lawfare ever seen, and so on—the thirteen prior years of MAS governance were put into even sharper relief.

The period 2006–2019 was now associated with *relative* political stability, indisputable national economic growth, at least a rhetorical commitment to decolonization, and the articulation of a "post-neoliberal" political economy (in theory, if not in practice) that became iconic globally. In other words, as the October 2020 elections got closer, Bolivians began to realize that a distinction could be drawn between the complicated and polarizing figure of Evo Morales himself and the MAS as an instrument for a particular vision of governance, socioeconomics, and international relations.

In addition, the ABM approach confronted another difficulty with their strategy: the fact that the MAS ticket had been crafted to ensure that, under the circumstances, the party had its best chance to be returned to power. Although key sectors of MAS's traditional Indigenous, peasant, and union base had demanded different alignments of candidates, the exiled MAS leaders, including Morales, decided on Luis Arce and David Choquehuanca as the party's choices for president and vice president.

Choquehuanca was the preferred political leader of the country's Indigenous movements. He was one of the founders of MAS along with Morales, an ethnic Aymara from Omasuyos Province near Lake Titicaca, and the minister of foreign affairs during all three Morales administrations. Although Choquehuanca was not nearly as charismatic as Morales, his ethnic and class background resonated powerfully with the all-important sectors of the MAS coalition that likewise identified with the regions, labor histories, and cultural heritage of the Aymara heartland (which is located mostly in La Paz Department).

Luis Arce had a very different background from either Choquehuanca or Morales but one that arguably was more relevant for understanding the past, present, and future of Bolivia's lithium project. The middle-class son of teachers from La Paz, Arce received his early schooling during the dictatorships of the late 1960s and 1970s. He then studied accounting and economics before beginning his career with the Bolivian Central Bank (BCB); he took leave from the bank in 1996 to complete a master's degree in economics from the University of Warwick in the UK. Arce spent his entire non-political career rising through the ranks at the BCB during most of the neoliberal governments after the restoration of democracy in the country. He developed a reputation at the BCB as the consummate economic technocrat, one of the leading experts in Bolivia on questions of debt management, currency policy, and international trade.

However, while working at the BCB, Arce also taught economics at the Universidad Mayor de San Andrés (UMSA), the major public university in La Paz and one with a long history of radical faculty and student politics, especially in the social sciences. Arce eventually came to view state-controlled economic planning as a potential instrument of social transformation, something that could play an important role in overcoming the harmful social consequences of Bolivia's neoliberal policies. His profile was thus quite unique: one of Bolivia's leading public bankers and economists who had also become a critic of many of the hallmarks of the prevailing economic orthodoxy. As we have seen (chapter 2), Morales's 2005 presidential campaign called for the nationalization of the country's oil and gas industries in the wake of the Second Gas War. The formulation of this and other economic aspects of the MAS platform bore the imprint of the anti-neoliberal philosophy of Arce, who had been brought on during the campaign as an economic advisor.

After the historic MAS victory, Arce was appointed as the country's finance and then economics minister. With the exception of a period of medical leave to receive treatment for kidney cancer, Arce oversaw Bolivia's economy from 2006 to the 2019 coup. He was the main architect of one of the most significant economic transformations in Bolivian history, a period of economic growth, poverty reduction, macroeconomic stability (fueled by the country's stockpile of hard currency reserves), and redistribution, which took place through a new system of *bonos*, or social benefit payments. Yet compared to marquee MAS leaders like Morales and Vice President García Linera—and even compared to other government ministers over these years—Arce kept a very low profile while in office. Ever the technocrat, he worked to steer Bolivia's economy from behind the scenes.

He was nevertheless at the helm through all the important milestones of the lithium project, even while he was directing the more economically pivotal development of Bolivia's state-owned natural gas industry. For example, although Bolivia's energy minister at the time, Rafaél Alarcón Orihuela, and the first director of YLB, Juan Carlos Montenegro, were the main Bolivian negotiators of the 2018 YLB-ACISA contract, Arce was responsible for ensuring that the deal complied with MAS's wider political economic model, which, as we have seen, was most clearly defined by the 2013 Patriot Agenda 2025 (PA 25), a document that itself bears many of the markers of Arce's economic philosophy.

Although it wasn't obvious at the time, in retrospect, one can attribute the later transformation of the lithium project not only to a growing worry at the highest levels over the kinds of confounding materialities described in the first sections of this chapter but also to a turn toward a certain pragmatism amid the country's continuing doctrinaire commitment to a set of political economic principles. In this sense, the emergence of flexible extractivism within the lithium project must be understood as both a response to a range of material and productive difficulties *and* the application of a political economic agenda almost hidden among a set of bold statements about the relationship between Bolivia's economic future and its pursuit of what PA 25 describes as "dignity and sovereignty."

With yet another convincing electoral victory for MAS in sight, Arce and his team began work on a "mid-range" development plan, one that would give more substance to the country's evolving political economic approach, while at the same time reaffirming its ideological adherence to the mandate of PA 25.[26] Released in early 2021, the "Economic and Social Development Plan 2021–2025" leaves no doubt about the importance of lithium industrialization for Bolivia's future. Indeed, the photograph chosen to introduce the plan shows the facilities on the salar, including the evaporation ponds, the industrial potassium chloride plant, and the industrial lithium carbonate plant (which was still under construction). After this introductory section, the plan then turns to a long section that "diagnoses" Bolivia's economic strategy in detail; again, the full-page photographs selected to represent this strategy are images of the multicolored evaporation ponds and, somewhat later in the document, an YLB worker loading salt into a wheelbarrow.

Using the year 2020 as the baseline, the plan describes a range of production and financial targets for the lithium project. For example, while shifting the estimated date for the completion of the industrial lithium carbonate plant to 2022, the plan makes a number of projections for 2025. Not only is this the last year of the five-year plan, but it is also a date with wider symbolic meaning: it is the year of Bolivia's bicentennial, the coming two-hundred-year marker that was also chosen as the framing device for the 2013 PA 25. But if the idea was to use the economic and social development plan as an instrument for a broader national historical reckoning, in practice it became a mechanism for reckoning more specifically with the lithium project. In a sense, the unfolding process of lithium industrialization became a litmus test for gauging the country's progress more generally, a unique form of pressure that was felt keenly by workers across YLB, from the technicians at the salar, to the project managers and policymakers at headquarters in La Paz.

But even without Bolivia's bicentennial looming just over the horizon, the projections for 2025 in the plan are extremely daunting. If by 2020, Bolivia had only managed to produce 944 tons of lithium carbonate, it was now projecting a cumulative total of 80,959 tons by the end of 2025. If total earnings from the lithium project were only $5 million to date, the government was projecting earnings of more than $3 billion by 2025. And although there were obviously no "surplus" revenues by 2021, considering the total debt owed by YLB to the BCB since the first credit was issued in 2008 (see chapter 2), the plan calls for the creation of a Norwegian-style sovereign wealth fund by 2025 with the surplus earnings from the lithium project. Finally, and most implausibly, the plan incorporates a projection based on the country's interest in Direct Lithium Extraction (DLE), even though at the time (early 2021) industrial DLE was still very much a technological abstraction. Nevertheless, the plan projects that *90 percent* of the lithium carbonate in Bolivia will be produced through DLE by 2025.[27]

With these wildly ambitious—one might even say chimerical—targets in place for the lithium project, targets infused with the urgency of broader national historical reckoning, the task then fell to YLB to square a circle. YLB had to demonstrate real movement toward these different goals while at the same time hewing to a number of stated political economic values like "productive sovereignty." Yet it was precisely this tension that opened a space for a set of decisions and adaptations that eventually stretched the existing framework for the lithium project to the point where the circle itself was broken—and then remade.

The first signs of the coming transformation were institutional. In 1970, during the military government of Alfredo Ovando, a new ministry of energy and hydrocarbons was established under the leadership of the socialist political leader Marcelo Quiroga Santa Cruz (who a decade later would be abducted and killed in the early hours of the Cocaine Coup of 1980). This ministry, which oversaw the hydrocarbon industry and electricity production in Bolivia, was maintained for almost fifty years.

But in 2017, the old ministry was dissolved. In its place, the government created new stand-alone ministries of energy and hydrocarbons in order to emphasize the distinctiveness between Bolivia's energy strategy and its robust gas and oil sectors. Even more, the new ministry of energy was subdivided into vice ministries: one to manage the country's electricity production and the other to oversee the development of "advanced technologies," which included just two—nuclear power and lithium.

However, this new high-profile ministry of energy lasted only three years. In November 2020, in one of its first major policy decisions, the new Arce government eliminated the ministry of energy by recombining it with the ministry of hydrocarbons, creating a new/old ministry in which "hydrocarbons" tellingly comes before "energy." The significance of this reorganization would soon become apparent at all levels of Bolivia's lithium project, including at the level of the key personnel's experiences. Yet at the same time that the Arce administration was laying the groundwork for broader shifts in orientation, it confronted the more immediate task of reestablishing the basic operations of YLB after the disastrous prior year.

The first priority was finding a new president for YLB, someone who would be both capable and willing to oversee the coming changes to the lithium project. But given the way in which the selection of YLB leadership had been an unrelenting three-year history of turnover and political conflict, this decision was anything but straightforward. Although the Arce government did its best to select a candidate with impeccable scientific and management credentials—that is, someone who wasn't first and foremost chosen based on political connections—it would soon become clear that the *absence* of such connections was going to doom the new YLB president.

Marcelo Gonzales had spent much of his early career as a professor of physics at Potosí's Universidad Autónoma Tomás Frías (UATF), during which he received a fellowship from the Federal University of Rio de Janeiro to complete a PhD in condensed matter physics. (A PhD is not a prerequisite for teaching jobs in Bolivian universities.)

After finishing his PhD in Brazil in only two years—he studied the movement of ions within the crystallized structure of lithium-ion batteries—Gonzales stayed on to pursue postdoctoral research on battery cathode synthesis. But with the lithium project moving forward in Bolivia, he was offered a position as one of the first Bolivian scientists at a new battery research and development facility in La Palca, on the outskirts of Potosí (see chapter 4). He moved back to Bolivia to take up the position and quickly rose through the ranks. By January 2021, he was serving as the director of the facility when he got a surprise phone call from La Paz while on holidays with his family at his house in Potosí, informing him that he had been named the new president of YLB. He was at the time only forty years old.

Despite the fact that he was only given a day to accept the offer, his decision was never in doubt. As he told me, *these kinds of opportunities don't come up very often, but I was still shocked by the phone call.* Nevertheless, he had little time to prepare: Gonzales was required in La Paz at the beginning of the following week for the public announcement, which had already been planned. The following weeks were a blur for Gonzales. He moved to La Paz and immediately took charge of YLB. As he described it, he was on "autopilot" for much of this early period.

On the one hand, he had to oversee a process of top-to-bottom restaffing in both La Paz and at the salar; do something about the evaporation ponds and industrial equipment, which had been damaged both by active sabotage and passive abandonment; and ensure that the construction of the industrial lithium carbonate plant resumed as soon as possible. On the other hand, Gonzales had to take command of YLB precisely at the moment in which the vision for the lithium project was in the process of being altered by those who had appointed him. In the end, these pressures proved to be too great for the first head of YLB under the Arce government, and the bookish physicist was forced to resign after only seven months in office.

The problem, as he put it, was that he approached these different challenges from a technical and scientific perspective. He tried to design a strategy for YLB that focused on quality control, the training of YLB workers, and research capacity. But what he didn't know at the time he was appointed was that the Arce government had already decided on a very different strategy for the lithium project, one that radically condensed the entire present and future of the project into one overriding objective: the industrial production of lithium carbonate in the shortest amount of time possible.

In addition, this largely unstated yet historic shift was taking place against a backdrop in which a growing understanding of the material complexities at the salar increasingly shaped lithium policymaking. All of this meant that the careful deliberations and the strategy put into place by Gonzales and his team, which should have been welcomed after the catastrophic year of 2019–2020, turned out to be distinctly *unwelcome*, the wider implications of which became crystal clear to the outgoing YLB president.

What he was witnessing was in fact the *hydrocarbonization* of Bolivia's lithium project. As he explained,

> *As you know, before there were two ministries . . . but with this merger, the vision [for] lithium has become more of a hydrocarbon vision . . . I became like a stone in the shoe because I wanted to focus on technical personnel, different long-term projects. But remember, I had come from the research center [in La Palca] and this is something they* [the heads of the merged ministry, but also perhaps Arce himself] *just couldn't wrap their minds around, the idea of researchers at Llipi* [site of the lithium facilities].

And then he put his finger directly on the inspiration for the major shift in the lithium project.

> *This is a hydrocarbon vision* [visión petrolera]*, because the new bosses come from Santa Cruz.*[28] *Do you know YPFB* [Bolivia's state oil and gas company]*? That is the model that is being used now [for the lithium project]. It's true that YPFB is a state company, like YLB, but they don't have a research center, they don't try and develop [Bolivian scientific] capacity. For YPFB, the only thing that matters is to produce. It's like making shoes, you just produce and produce.*

If this is true, I replied, what does it mean for the wider lithium project, the government's long-term ambitions?

> *If this happens with YLB as well, what this means is that everything will be reduced to the extractive part* [la parte extractiva] *and in the end we will only sell the raw materials, like we have always done in our history . . . For me this will be the end of the dream of creating our own productive chain from the extraction of raw materials to the manufacture of lithium-ion batteries.*

With the ouster of Marcelo Gonzales, which was followed by yet another process of widespread housecleaning of personnel at YLB, especially at the levels of policy implementation and basic research, the hydrocarbonization of the lithium project began in earnest. First, the geological, environmental, and productive complexities at the salar became increasingly part of the discussions around the future of the project. Salvador Beltrán, YLB's forty-year-old director of investigations, engineering, and development, was quite explicit about how much the wider lithium project had been rethought in a relatively short amount of time as the true nature of the material challenges at the salar became impossible to ignore.

He emphasized three main issues: the high levels of magnesium in the salar's brine, which complicate the process of purification needed to produce battery-grade lithium carbonate; the ways in which rainfall at the salar makes brine evaporation a highly inefficient mode of extraction; and the specific structure of the salar's geology near the southern and southeastern zones— the zones of greatest importance for YLB—where the salar's strata are relatively shallow, which means that the brine that is pumped out must be replenished through natural processes. As he explained,

> Unlike other places [especially Chile], our salares depend a lot on the hydrological recovery time of our aquifers, so although we have a gigantic reserve [of brine], if we don't manage the water resources that nourish the brine with lithium, we might actually destroy the salar, which would mean of course that production would stop.

YLB's director of operations, Karla Calderón Dávalos, was thinking even more expansively about hydrological problems during this period. Calderón had risen through the ranks of the state lithium company very quickly, in large part due to her expertise as a systems engineer with specific training in "evaporitic resources" and a well-defined vision of the historic importance of the lithium project for the country's future. A proud potosina, she was one of the top students at the city's Universidad Autónoma Tomás Frías (UATF). In 2015, she was the only UATF graduate to receive a highly competitive study-abroad grant from Bolivia's state mining company COMIBOL, which, to recall, directed the lithium project until YLB was created in 2017 (see chapter 2).

Calderón decided to pursue a master's degree in France, at the Université Grenoble Alpes in the southeastern city of Grenoble, one of the country's main centers for scientific research outside of Paris. She spent two years in Grenoble, a time of intensive study and network-building that had a profound impact on her scientific outlook and sense of purpose. Obligated by the terms of the government grant to work for a Bolivian state company upon her return, she eagerly embraced the opportunity to be part of the cohort of young Bolivian scientists charged with implementing Bolivia's emerging lithium strategy. As Calderón put it, the country "was expecting great things from us."[29]

But when I met the thirty-one-year-old Calderón in 2022, she was agonizing over the problem of water, specifically the water that would be necessary for the much-delayed industrial lithium carbonate plant to function. As she explained, the industrial lithium plant will require an enormous amount of water, which will be used primarily for a massive boiler system. The boilers will produce steam, which will maintain the industrial machinery at the temperatures required to produce lithium carbonate.

When I asked her how much water would be needed for the industrial plant, Calderón's expression belied a weariness about an intractable problem that has been an enduring source of conflict for lithium producers in South America, in which industrial water usage has profoundly impacted communities near the zones of extraction. *We are talking about approximately three hundred liters per second*. Three hundred liters per second? I replied, trying to fathom how this amount translated into liters of water per year. Seeing my surprise, Calderón simply said, "Yes, that's a lot of water."

But what is the source for these unimaginable quantities of water? The only river in the area is the Río Grande, whose water provides sustenance to plants and animals and the town of Río Grande itself.[30] And the water from wells connected to this river would be used in the operations of the industrial lithium plant. But as Calderón explained, the water in the Río Grande is too "heavy" to use directly, given the river's tight connection with the hydrology of the salar. This means that the river water must first be treated, also at an industrial scale: it must be "softened," which removes different impurities— especially salt—that would be corrosive for the equipment. Softening also lowers the conductivity of the water.

At this point, I became confused. I had been exploring the delays and complications around the construction of the industrial lithium plant for years by this stage in my research, but suddenly, an entirely new project came into view: an industrial water treatment plant, which, as Calderón acknowledged, will be by far the largest such facility in Bolivia. Furthermore, the lithium plant, which itself hadn't been completed, could not become operational until the companion water treatment plant came online. And construction on this second plant hadn't even begun. Ever the professional, however, Calderón didn't belie any unease over this major obstacle. As she explained, YLB was developing a contingency plan that would allow the completed industrial lithium plant to begin "start-up tests" once the treatment plant was producing even limited amounts of purified water.

At the level of policy, all of these factors and more confronted YLB with a serious dilemma: how to shift as quickly as possible to the industrial production of lithium carbonate at the same time that the material and infrastructural realities at the salar made such a shift so difficult—regardless of the urgency of the messages coming from La Paz. Yet if the material complexities of lithium production shaped the transformation of lithium policymaking, so too did another distinct dimension: the legacy of history.

The historical reckoning at the heart of the lithium project involved a number of strands: the weight of longstanding international skepticism about Bolivia's capacity to finally extract lithium at an industrial scale; the recognition within the Arce government that abstract ideological ambitions like "productive sovereignty" had not always served MAS's process of change well, especially as the country's gas reserves continued to diminish; and, as we have seen, the fact that the coming bicentennial in 2025 exerted pressure at all levels as an inescapable and critical inflection point. But if YLB officials like Beltrán were left to respond at the level of policy to the ineluctable actualities of the lithium project, it fell to the head of YLB himself to carry the burden of history.

In August 2021, in a move that also served as one of the first unambiguous institutional signals of the hydrocarbonization of the lithium project, Carlos Humberto Ramos Mamani was named the new president of YLB. The potosino Ramos was an electrical engineer who had spent most of his career working in Bolivia's mining, oil, and gas industries, including with the state hydrocarbon company YPFB. In particular, he had specialized in the management of industrial plants and large-scale extraction sites. Yet, like so many key figures in the unfolding saga of lithium in Bolivia, he came to his post with no expertise in the diverse particularities of lithium production.

But even if Ramos didn't have a background in "evaporitic resources," he had more than enough experience working in Bolivia's hydrocarbon and mining sectors to know that he had been given an impossible mission, which included the intense pressure coming from all sides: from the frontline operations at the salar; from frustrated personnel within a YLB headquarters that was under the constant gaze of scrutiny; from the newly installed ministry of hydrocarbons and energy, which was itself under heavy pressure; and from the Arce administration, which had put almost all of its chips on lithium as the resource that would replace gas at the foundation of the country's economy.

Because Marcelo Gonzales had lasted less than a year, it fell to Ramos to make the official case for the key objectives in the 2021 economic and social development plan, all of which pointed to 2025 as the moment of reckoning. And while he was more than willing to at least *appear* to take these objectives seriously—that was his job, after all—he viewed the situation with an affable realism tempered by a kind of gallows humor.

By August 2022, one full year into his tenure as head of YLB, Ramos knew very well that the chances of meeting *any* of the government's lithium targets for 2025 were close to nil. At the same time, he also knew that, like so many before him, a Sword of Damocles hung over his head, that he was likely destined to join a long line of scapegoats for the complex of factors that have made Bolivia's lithium project one of the most fraught undertakings in global economic history.

I told Ramos that I had learned through a variety of discussions with engineers and technicians, both at the salar and elsewhere, that the 2025 goals for lithium production were impossible to meet under current conditions. For example, the idea that Bolivia will have produced almost eighty-one thousand tons of battery-grade lithium carbonate by 2025, which would mean Bolivia would go from being a *non*-industrial producer to the second largest producer of lithium carbonate in the world (after Australia), was sheer fantasy. Maybe, just maybe, by 2035, but surely not in only three short years . . .

> *No, I understand your question, but the date for us is 2025, not 2027, not 2028, but 2025. I know those are big volumes, and I know that pressure is being felt, but it's not like they tell me, "Tomorrow we want something to start up." No, we have a plan, and we are just following the plan.*

But do the plans actually provide a way for YLB to begin producing lithium carbonate at these levels by 2025? Ramos then revealed the extent to which he fully realized that he was trapped in a paradox.

> *Well, yes and no, it is not going to be achieved, but we have to find a way to reach these values anyway, so at this moment we are working on planning, step by step.*

But what if it is not technically possible to begin producing lithium at the necessary levels to come anywhere near the 2025 targets?

> *I know, but my number is 2025. Those who have to do something, do it.*

And then Ramos went off the official script as head of YLB. He had probably grown tired of the charade of having to defend or explain his impossible situation and wanted, instead, to reflect more generally on how he viewed the *real* future of the country's lithium project. It was a vision that was much more ambiguous but still supportive, even optimistic.

> *Look, I do think that there will be a time in which YLB will eventually produce lithium and will obtain a return. The challenge is going to be how to maintain and increase those production volumes, because we understand that it is a state-owned company and sometimes it is a little bit difficult to move forward as a*

state company, because there are many processes, they are very demanding and you have to have very strong justifications to make investments, right? But we know, it is lithium, it has to happen, it has to be produced . . .

CONCLUSION: ENERGY TRANSITION AND THE PRODUCTIVE IMPERATIVE

The ethnography of Bolivia's lithium project after the return to power of the MAS under the leadership of Luis Arce opens a window into broader shifts in the country's political economy, shifts that hold important lessons for the future of the wider energy transition. What happens when a country's reliance on resource extraction as the basis for social and economic growth, including meaningful redistribution, is tied to a resource that is part of the supposed global solution to the climate crisis? Although lithium is not a renewable resource like wind or solar power, its use in lithium-ion batteries makes it one of the few essential resources in the international strategy of decarbonization. And what does it mean that the extraction and commodification of a decarbonizing resource like lithium hold the potential to bring vast amounts of *public* revenue to a country like Bolivia, which has been governed since 2006—with the exception of the post-coup year—by a political party and movement that is committed to a complex yet undeniably transformative "process of change"?

The evolution of a political economic framework for the lithium project in Bolivia that reimagines the industrialization of lithium as a pillar of *both* national development *and* the global energy transition shows that the answers to these and related questions are being debated well beyond the centers of international policymaking and critique. But what are the wider implications for the future of the energy transition of the move to rearticulate the framework for lithium industrialization in Bolivia, a move that both responds to and anticipates actually existing complexities across the entire productive process, from brine evaporation—and, potentially, "direct lithium extraction" (see chapter 6)—to the eventual sale of thousands and then tens of thousands of tons of battery-grade lithium carbonate?

On the one hand, the insistence on a new framework in Bolivia for lithium extraction—a framework that is deeply intertwined with international climate policy—arguably undermines conventional understandings of both extractivism and neoextractivism, especially since much of the critical literature focuses on the destructive impacts of the mining, oil, and gas industries. In this sense, it is important to note that the hydrocarbonization of the lithium project in Bolivia is a state-response to the range of complexities and forms of resistance analyzed in this chapter and *not* a signal that the state has given up on its long-term plan to replace gas and oil with a resource at the center of an imagined postcarbon future—quite the contrary.

Yet on the other hand, as with the shift to more flexible modes of accumulation within capitalism, the development of flexible extractivism in Bolivia and elsewhere, particularly across the resource topographies of the global South not only does little to challenge the underlying economic logics of the energy transition, it actually renders these logics that much more immovable.[31] To the extent that the energy transition depends on the replacement of fossil fuels with supposedly more sustainable forms of power, including lithium-ion batteries, this long-term transition *will* take place through the mechanisms of resource extraction, industrial production, global trade, and mass consumption. In other words, the ethnography of flexible extractivism in Bolivia is also a window into not only the future of the energy transition but also the future course of climate crisis mitigation and other efforts to counteract the "slow violence" of planetary deterioration.[32]

And finally, the recasting of the lithium project in Bolivia through an emerging political economy of flexible extractivism holds lessons for broader debates over the shape and form of decolonial unmaking and the ultimate possibilities for self-determination within an Anthropocene infused with inequality, enduring injustice, and ecological trauma. In much the same way that insurgent forms of transnationality in the early post–Cold War period challenged the traditional hierarchies of both international politics and cultural identities without overturning them,[33] the move toward flexible extractivism likewise offers both a challenge and rebuke, especially of visions for "climate justice" that have very little to say about the socioeconomic needs of states and communities whose choices are constrained by forces of political history, geography, and economic vulnerability.

This is not to say that states like Bolivia—not to mention communities throughout Bolivia—are monolithic in any way; far from it. As the critical geographer Penelope Anthias has argued, communities in regions at the center of the country's historic—if waning—gas and oil economies deploy what she describes as "pluri-extractivism," in which the legal category of "autonomía" is mobilized by local leaders to make claims over gas royalties rather than as a strategy to resist gas production itself.[34]

But even here, the relationship between what might be thought of as the productive imperative and local visions of self-determination and autonomy is deeply contingent. In other words, well beyond the impositions of the state, many communities in Bolivia—and, again, elsewhere—themselves adhere to a version of flexible extractivism, both in relation to the lithium project and otherwise.[35] This is yet another reminder of how preexisting assumptions about the "assemblage of disparate elements, practices, and processes that constitute actually existing socioeconomic life are often profoundly misbegotten.[36]

FOUR

The Cathode Chronicles

IN 2019, AFTER MONTHS of preparation and planning, my much-anticipated first visit to the headquarters of YLB in La Paz began somewhat inaus-
piciously: despite my best efforts and knowledge of the city and its incomparable topography, I had a very difficult time even finding it. Years before, I
had spent time conducting research for a different project at the headquarters
of the state mining company, COMIBOL, which is located in the very center
of the city, on Avenida Camacho, known as the "Wall Street of La Paz." This
boulevard was part of an urban project that began in the early part of the
twentieth century to concentrate the country's main financial institutions
along the same city block (it also later became the place where people went to
buy and sell currency—mostly US dollars—from informal street traders).
After the creation of COMIBOL in 1952, it was housed in a building with a
neoclassical façade, whose architecture was meant to rival the neighboring
buildings of finance and trade. The corner of the COMIBOL building,
where Avenida Camacho intersects with Calle Loayza, features a unique
design element: an inlaid balcony, at street level, in front of a granite wall on
which the COMIBOL logo is affixed. This logo famously features an etching
of a Bolivian miner's face, with hollow eyes and a contorted expression that
evokes as much suffering as resilience.

Given that Bolivia's lithium project had been directed by COMIBOL
until 2017, I had assumed that the offices of YLB would be either in the same
building or at least in another of the many aging state-owned structures in
central La Paz. However, I soon learned that YLB was not where I had
expected it to be. Instead, the Morales government had decided to give the
lithium project an institutional location that was meant to signal its unmistakable importance for the country's future. As far back as 2014, three years

before the establishment of YLB as Bolivia's state lithium company, the government had decided that the project would be directed from the lofty heights of Edificio Hansa, the iconic Juan Carlos Calderón–designed skyscraper just above the city center. Eager to present myself to YLB officials, I rushed up the Prado, the principal street of La Paz, and entered the lobby of the imposing glass landmark.

For moments like this, I have developed a particular ethnographic technique, one that has served me well over two decades of research in Bolivia, in which I try and draw as little attention to myself as possible. It's not that I attempt to hide or disguise my immediate purpose, if asked. It's just that I have learned through long experience that the always-fraught process of ethnography can be smoothed, if only somewhat, by having a destination, an objective, a prearranged appointment, an institutional contact person—or, at least, by having the appearance of such various certainties. To flounder around, to look confused or lost, to show frustration is to invite scrutiny, delay, the need for what can become a lengthy explanation, in ways that complicate an already unpredictable methodology.

So as soon as I was inside the building, I went immediately to the tenant directory; my hope was that I would quickly find the entry for YLB and would, in a matter of seconds, be in the elevator ascending to its offices. I stood in front of the large glass cabinet, scanning the names, beginning with the first floors. The building was filled with different private businesses: law firms, financial consultants, and telecommunications companies. The only tenant of Edificio Hansa that seemed even remotely connected to YLB was the Federación de Cooperativas Mineras Auríferas del Norte de La Paz, the union that represents cooperative gold ore miners in La Paz Department. After looking through the list several times, I was certain that YLB was not among the tenants. In irritated bewilderment, I left the building and looked straight up. There it was: "Hansa," written right above the entrance. Could there be more than one Edificio Hansa in La Paz?

I ducked back into the lobby and gave the tenant directory one more perusal. Maybe the altitude was playing tricks with my mind? I had only just arrived the day before and was finding that, in general, I needed more time to acclimatize to the thin air the older I became. My strong urge to rely on ethnographic efficiency was now replaced by disorientation; I had no choice but to approach the building's concierge, who was sitting quietly in a booth a few meters from me. Excuse me, I said, I'm looking for *Yacimientos de Litio Bolivianos* (I used the full name, since many Bolivians hadn't heard of

"YLB"). The concierge looked up and I fully expected him to put in motion a variety of gate-keeping procedures that would require me to provide the kind of information I hoped to obtain *during* my first visit with YLB officials. But with luck finally on my side, he simply said, *YLB is on the nineteenth floor.*

I thanked him and turned away, but before heading to the elevators, I went back to the tenant directory and looked one more time. The headquarters of YLB are on the nineteenth floor? And there it was: the listing for the nineteenth floor on the left and, on the right, next to it . . . nothing. The entry for the nineteenth floor was empty, the only officially vacant floor in the entire building. What did this telling absence mean? Was the Morales government trying to keep the actual location of Bolivia's new state lithium company a secret, or at least partially hidden, for some reason? Had YLB received threats? And if so, from whom? With these wild thoughts swirling, I took the elevator to the upper floors and found my way to a door whose brass plaque announced that I had, indeed, come to the right place.

I didn't ask about the lack of signage in the lobby but instead launched the process I knew so well: the short explanation for my visit to the YLB receptionist, who sat next to a uniformed police officer; the delivery of a letter of presentation addressed to YLB's current president, which contained a much longer description of the research project and request for institutional collaboration; and, these initial formalities out of the way, the more informal request to the receptionist to see if anyone was available for a quick chat *now*, given that I was already there. In my experience, a government receptionist at this moment will usually demur with a response along the lines of "We have your cell phone number and email, and someone will be in contact." But sometimes, as now, the receptionist will pick up the phone and dial someone, usually, in cases like this, someone in the communications department, which is responsible for frontline contact with a wide spectrum of interested outsiders, from journalists to school groups to anthropologists.

The receptionist told me that someone would be with me soon and asked me to take a seat in the small waiting area directly across from her desk. As I sat there, I watched an YLB video on a television screen above the head of the police officer, a piece of company propaganda that played on a two-minute loop. The sound was off, but the images were of the various lithium facilities at the salar, interspersed with smiling workers and shots of YLB dump trucks loaded with salt rumbling toward the industrial plant. But in addition to the installations at the salar, the short video contained a segment from another facility, one I didn't recognize; there were what looked like advanced laboratories with workers in the kind of protective clothing that I associated with hospitals, not lithium extraction. I was still contemplating this mystery facility when a young woman entered the waiting area and asked me to follow her.

After briefly explaining the reason for my visit—the fact that I was at the beginning of a four-year research project on the country's lithium process and would appreciate any assistance that YLB was willing to offer—she cut me off somewhat abruptly: *From what I understand, you have already given us your letter [carta de solicitud]?* Yes, I replied. *As my colleague mentioned, we will be in contact with you after it has been considered by the president's office.* Very well, I said, realizing that it was going to be a short meeting. As I stood up, she stopped me: *We'd like to give you a gift, to thank you for your visit.*

She then handed me a small, white plastic object with a stylized YLB logo stamped on one of its sides. It was oval-shaped and thin. I held it in my hand, not knowing what I was supposed to do with it, or what it was. She told me to squeeze it, which I did, and almost blinded myself in one eye when it emitted a laser beam, which I unintentionally had turned toward my face. I quickly moved it in the other direction while she explained, *it's a laser pointer, for meetings, with an YLB lithium-ion battery inside of it.* I didn't know that YLB was actually *producing* lithium-ion batteries until that moment, despite the fact that Bolivia's lithium "master plan" involved eventually manufacturing batteries and even EVs—the final steps in the country's ambition to complete the lithium "productive chain."

In other words, although I was fully prepared to study the various dimensions of lithium industrialization, which itself seemed like a distant endpoint in a distant future, I realized that the lithium project in Bolivia now encompassed another facet: YLB, the country's state lithium company, was also now in the business of manufacturing batteries, even if the initiative was still very much incipient. The laser pointers, for example, which worked beautifully as far as laser pointers go (and were designed with a small USB recharging port), contained a small battery with a relatively low capacity (3.7 volts and 8 amp-hours [Ah]).

But as I would later learn, the cathode, or positive electrode, in the battery was made with lithium cobalt oxide (or LCO), a compound that *wasn't* produced by YLB and wouldn't feature in the "production lines" that would be developed in-house over the coming years. Still, the fact that YLB had developed the capability by 2019 to assemble a small lithium-ion battery from existing components and produce an electronic device that used this battery, was nevertheless impressive. Although I had become transfixed, like so many others, by Bolivia's drive to industrialize the world's largest reserves of lithium, it was clear that the battery link in the chain was—at least at the time—becoming an important part of the wider strategy. But how important? As I held the laser pointer in my hand while descending back into the streets of La Paz, a number of questions came to me.

How serious was the government's plan to extend the lithium value chain from lithium extraction to *at least* the production of "high capacity" lithium-ion batteries, not only at a "pilot" level, but at industrial scale? Very few countries in the world had even attempted to interconnect the industrial production of battery-grade lithium carbonate with the industrial manufacture of lithium-ion batteries, especially those destined for use in the battery packs that power EVs. For example, the country whose mines produce the most lithium carbonate, Australia, doesn't have a lithium-ion battery industry. The same is true for Chile and Argentina. And countries like South Korea and Japan, whose private companies are among the leading producers of lithium-ion batteries, don't have any appreciable lithium reserves.

The only country in the world to produce both lithium carbonate and high-capacity lithium-ion batteries at a global scale is China, which has about 4.5 million tons of lithium reserves and whose companies dominate the lithium-ion battery market (including CATL, which supplies about 37% of all lithium-ion batteries worldwide). And China is unique in yet a different sense: its companies hold controlling or partial stakes in some of the largest lithium operations in the world, including the Greenbushes mine in Western Australia (co-owned by Tianqi Lithium), Chile's SQM (also partly owned by Tianqi), and the Cauchari-Olaroz mines (on two adjacent salares) in Argentina (co-owned by Ganfeng Lithium). This means that China produces lithium carbonate from its own reserves—which are found in brine in western China (on the Qinghai and Tibetan plateaus), in spodumene rock in Sichuan province, and in lepidolite rock in the country's northeast near the border with Russia—while it supplies its massive battery industry from mining operations it controls beyond the country's borders.[1]

So when viewed through the lens of the global lithium energy assemblage, the mounting pressure on Bolivia and YLB to begin producing battery-grade lithium carbonate at industrial levels makes perfect sense. China might dominate the entire lithium energy assemblage, which includes, importantly, the manufacture of EVs (the Chinese EV company BYD is now the largest in the world), but the future of the green energy transition still depends on

Bolivia's lithium. As we have seen, without Bolivia's lithium, there simply won't be enough remaining reserves in the world to meet the astronomical projected demand for lithium-ion batteries over the coming decades. And without Bolivia's lithium, the price of batteries and thus of EVs—not to mention electricity storage systems and other applications—will continue to be prohibitive for most people in most countries.[2] Without Bolivia's lithium, in other words, a green energy transition based on the replacement of fossil fuels with lithium-based energy will remain either a fantasy, or the exclusive purview of the elites who can afford it, or both.

Yet when viewed through this same global perspective, the angle that helps explain both a wider lithium geopolitics and the evolution of Bolivia's national lithium project, the effort to establish a lithium-ion battery industry in Bolivia is perplexing. Given the conventions of supply chains, labor costs, economies of scale, and the other well-established structuring logics of global capitalism, there are good reasons why Chinese, South Korean, and Japanese companies control the lithium-ion battery market. They are the same reasons why the end-users of batteries, especially EV automakers, have so far been content to simply buy Chinese or South Korean batteries rather than build the industrial capacity to make lithium-ion batteries themselves. From this perspective, the idea of developing a battery industry in Bolivia that could upend these alignments makes no sense.

But as will be seen, the push to develop the capacity to produce Bolivian-made batteries using Bolivian-made lithium carbonate—while necessarily importing other components—must be understood differently, less through the imposed demands of the wider lithium energy assemblage and much more through the nuances of political ideology rendered into a strategy of scientific nationalism. Furthermore, the ethnography of Bolivia's battery gambit reveals worlds of expertise and production that exist at smaller, even intimate scales, many of which are completely removed from the pressures of the global battery market. These are worlds that are bound up with the lives and professional aspirations of young Bolivian researchers, whose participation in what will likely be, in retrospect, a short-lived experiment within the much longer-term state lithium project has meaning that is both personal and transformative.

At the same time, if the accumulated weight of material, historical, and geopolitical complexities led the Bolivian government and YLB to eventually shift its approach to lithium industrialization, a shift that I describe and explain through the categories of flexible extractivism and hydrocarbonization (see chapter 3), something similar took place within YLB's "electro-chemical and battery" operations. However, the differences between these two corresponding shifts are critical: if flexible extractivism and hydrocar-bonization meant a paring away of many of the ideological imperatives that had originally undergirded the lithium project, in favor of industrial produc-tion at all costs, the changes to battery research and assembly were quite different, if no less consequential.

In particular, as the initial plans to develop a national lithium-ion battery industry in Bolivia faded relatively quickly, for reasons that will be explored below, researchers continued experimenting with the production of noncom-mercial devices, including those destined for use within Bolivia as part of the government's wider social development program. As will be seen, these alter-native lithium-ion battery projects were designed on the margins of the offi-cial, and ultimately doomed, initiative to introduce high-capacity Bolivian batteries into the global market. Yet these small-scale projects formed the technological basis for something like a real energy transition, even if it was a transition limited to energy use and storage within Bolivia's underserved rural communities.

Nevertheless, the story of the largely unacknowledged effort to repurpose Bolivian batteries as a form of sociotechnical redistribution, an effort that was made possible through the growing capacity of Bolivian battery scientists and technicians, offers lessons for how the wider green energy transition might likewise continue to unfold at unseen and local scales in ways that come to approximate—even if only partially—the ideals of energy and cli-mate justice that are proving so elusive globally.

· · ·

This chapter examines the trajectory of Bolivia's battery ambitions, which evolved in the shadows of the wider lithium project. In the next two sections, I situate the battery operations in relation to a number of different regional, political, and technological contexts. As will be seen, the geographical loca-tion of the battery "industrial complex" itself was as perplexing as the idea that this unlikely place would become ground zero for Latin America's first lithium-ion battery industry.

I then trace the contours of the physical, social, and affective spheres of Bolivia's battery research, development, and production facilities, which include the lives and aspirations of the scientists and researchers who were charged with rendering a utopian vision into the material terms of battery chemistry, electrical engineering, and component assembly. My research among these remarkable spheres examined the ways in which the scientific status of these facilities, as well as the expertise of its workers, slowly emerged as the facilities themselves became operational.

This chapter explores the nature and meaning of this emergence rather than treating the battery facilities as a (field) site where well-defined forms of knowledge were reproduced under a given set of conditions. In this sense, my approach to the Bolivian battery project and, in particular, to the ways in which its personnel became embodied agents of statecraft while also reconfiguring the project in part *against* the demands of the state, evokes Cori Hayden's study of Mexico's insurgent generic drugs industry.[3] Like the generic drugs workers in Mexico, the YLB battery researchers and technicians were also compelled to work and foster their emerging professional identities at the intersections where social provision, technological struggle, and wider market forces collide.

After examining the ethnography of crystal analysis, cathode synthesis, and the social lives of electron microscopes, among others, I turn to the question of production. What does it mean to "produce" a lithium-ion battery? As will be seen, YLB personnel established what I describe as a form of "pilot Fordism": a method of production that paradoxically incorporated elements of assembly-line techniques in the spirit of what became the artisanal, rather than the mass, production of both batteries and electronic devices. At the same time, and with very little public notice, YLB workers at the battery facility began experimenting with alternative products that were intended to be distributed noncommercially in rural areas *within* Bolivia. These devices had nothing to do with the broader MAS strategy to "extend the productive chain" through the industrialization of *both* Bolivian lithium and Bolivian batteries, but they were viewed by YLB researchers as batteries and devices with concrete applications that could actually improve people's lives.

The chapter concludes with a final section that returns to the wider narrative arc of the book, viewing the relatively isolated—both geographical and technological—battery operations of YLB in relation to the much more prominent shifts that were taking place around lithium extraction, industrialization, and the willingness of the Bolivian government to enter into joint ventures with foreign companies. As I argue, the broader hydrocarbonization of the lithium project—the push toward production at all costs and the elision of the initial ideological framing of the project—had significant consequences for Bolivia's battery ambitions and researchers, who were subsequently marginalized despite the investments by the state and a deep sense of purpose among these battery scientists. As will be seen, the practical future of Bolivia's battery operations became clear by 2022, but the experiences and scientific capacity that were nurtured through this unlikely episode in Bolivia's wider lithium saga will no doubt endure.

"ENERGY THAT UNIFIES"

Once I realized that I would have to do whatever I could to peel back the layers of YLB's battery project, I went about laying the groundwork for what became a multiyear encounter with the people and scientific culture of one of the more extraordinary initiatives I had ever seen in Bolivia. This meant first reacquainting myself with the city of Potosí, which became a hub for my research on the battery side of the wider lithium project, a hub that came to encompass lithium researchers and advisers at the Universidad Autónoma Tomás Frías (UATF); YLB personnel who worked at the nearby battery facilities but who lived in the city and were therefore more likely to be available for interviews outside their hours of work; and Potosí's different governmental institutions, which were obvious sites of keen interest given the history of anti-MAS and anti-lithium mobilizations in the city (see chapter 2).

Although I had a long personal history with Potosí Department, my own past experiences in the city of Potosí itself had been limited. At the beginning of my research career in Bolivia, I spent a formative year living and traveling throughout the north of the department, an iconic region known for its Indigenous social structures called *ayllus*. The norte de Potosí, where anthropologists like Olivia Harris and Tristan Platt had conducted groundbreaking research in the 1970s, consists of five provinces that are extremely isolated from the departmental capital far to the south. Indeed, most of the people I met during my time in the north, especially the majority Quechua- and Aymara-speaking agropastoralists who live outside of the small provincial towns, had never made the arduous and time-consuming journey to the departmental capital in their lives. Although I did spend several weeks conducting research in the Archivo Histórico de Potosí and the Casa Nacional de la Moneda, which contain an invaluable collection of documents and records from the colonial period, like my interlocutors from the north, I also had little reason to make the long journey to the city of Potosí.

But over the course of four years of research on lithium, I came to understand the city of Potosí and its complicated "civic" identity in a different light. Although much of the city's political and economic history extends through many centuries into the past, as we saw in chapter 1, a history of both extractivist splendor and human misery, it is still viewed by many potosinos today as the "city which has given most to the world and has the least."[4] With the long-since depleted Cerro Rico looming over them as an ever-present reminder of this history of rise and steep decline, it becomes much easier to understand why potosinos would look longingly toward the lithium project, despite its remoteness from them, as a potential source of economic and social rejuvenation.

Yet even here, the city of Potosí is likely fated to be left bereft. Although the Salar de Uyuni is almost entirely within the boundaries of Potosí Department, including, most importantly, the southern and southeastern zones slated for industrialization, the salar's lithium has been declared a *national* resource. This means that the department doesn't have any control over or legal rights to the lithium reserves within its territory. The city and department might someday *be granted* the right to receive a percentage of future royalties on the sale of lithium carbonate by the state—that is, beyond what would be redistributed through existing social programs—but even this remains very much in doubt.

All of this casts a particular light over the parallel, if much more opaque, battery project taking shape about thirty minutes from the center of Potosí, a rural location beyond the city limits that underscored its distance from the ambit of city and departmental governance. Unlike the Salar de Uyuni and its lithium, however, the battery facilities, while technically under the control of YLB, were not trapped within the legal and economic black box of a "fiscal reserve." Perhaps Potosí's political and business leaders might find a way to exercise more influence over how the battery project developed? But first, the city had to reconcile with national MAS leaders in the aftermath of almost a decade of strident opposition, including, most recently, the COMCIPO-led hunger strike and mobilization against the YLB-ACISA contract in the months leading up to the November 2019 coup d'état (see chapter 2).

Yet by 2022, the political landscape in Potosí had changed dramatically. With MAS resurgent, the antigovernment civic forces within the city were purged, including the arrest and detention of the former head of COMCIPO and failed vice-presidential candidate Marco Pumari, who was charged with committing various crimes during the coup. Pumari, who was only 38 years old at the time he was swept up in the national crisis and right-wing takeover, had returned to Potosí in utter humiliation soon after the resounding MAS victory in the October 2020 elections. His political career as a civic leader was over and his days as a free person were numbered.

With many in the city furious at Pumari for joining his lot to the failed political fortunes of Luis Camacho, the far-right militant from Santa Cruz, Pumari made a desperate, almost cinematic, gesture: before arriving in Potosí, he took to social media to invite all who wanted to confront him to meet him the next day in the city's historic Plaza 10 de Noviembre. As he put it, "I'll be at the Cathedral at high noon."[5] The next day, surrounded by a phalanx of police officers, Pumari was greeted by a large and angry crowd, which pelted him with "coins, oranges, and tomatoes."[6] He was arrested and imprisoned less than two months later.

Against this background, city and departmental leaders, many now (re-) affiliated with MAS, were eager to *finally* show their support for the wider national lithium project, as much for political survival as anything else. The new leadership signaled this strategic change of heart by announcing the creation of its own lithium office, whose stated mission was to do whatever it could to facilitate YLB's operations, both at the salar and, even more so, at the much closer battery facilities. At the same time, the department adopted a new slogan for itself that reinforced just how far Potosí had come since the recent era of anti-lithium protests: "The Autonomous Departmental Government of Potosí—Energy that Unifies."

The head of Potosí's new lithium office was Juan Téllez Rodríguez, yet another unsung protagonist in Bolivia's unfolding lithium project with a compelling life story and sense of purpose. Téllez had immigrated to Halifax, Nova Scotia, in the early 1990s to study for a second master's degree, after having earned his first, in development planning, from the University of North Carolina at Chapel Hill, where he was, as he put it jokingly, "a friend of Michael Jordan." Téllez then spent the next twenty years working and living in the maritime city of Halifax (motto: "From the Sea, Wealth"), where he married, raised a family, and became a Canadian citizen. His varied career in community development included working on Cape Breton Island to help its residents cope with the collapse of the cod fishing industry. Téllez was also a long-time adjunct professor of development studies at Saint Mary's University in Halifax.

However, in 2010, several months after Evo Morales and the MAS had been reelected in December 2009 with the largest majority in Bolivian political history (at least since the restoration of democracy in the early 1980s), Téllez was offered a position as an advisor in the national planning ministry. Although he had built a satisfying life in Halifax, both personally and professionally, he was pulled back to Bolivia at this historic moment when MAS's long-term process of change was given new institutional force and political legitimacy. I asked him if this was a difficult decision, to leave everything behind and return to Bolivia.

> *I love Canada a lot, it has given me so much. You can say that I have been very lucky in Canada, it's where I found my professional path. But I've always been tempted over the years to return to Bolivia, to make a wider contribution in life.*

After much reflection, Téllez accepted the post and returned to La Paz. As he explained, he had watched the rise of MAS and its "democratic and cultural revolution" with a powerful mix of pride and longing; as a well-educated Bolivian expatriate, he felt compelled to put his skills and training to use. Yet his work for the national MAS government soon left Téllez longing for the more academic dimensions of planning and development. In 2013, only three years after returning to Bolivia, Téllez left again, this time for Mexico, where he enrolled in a doctoral program as a mature student at the *Universidad Autónoma de Zacatecas*. After completing his mandatory courses in about a year and a half, he was able to return to Bolivia to conduct research and write his dissertation. In 2016, he was awarded a Ph.D. based on a project entitled "The Possibilities and Challenges of Living Well in Bolivia."

At the same time, however, Téllez's life took yet another major turn. While conducting research for his dissertation, he was contacted by friends and family members from the small town of Betanzos, located about forty-five kilometers from Potosí, where had grown up. After living most of his adult life in Canada, after receiving advanced degrees from universities in three different countries, and after working for the national government in La Paz, the "people of [his] community" had a message for him: *You've been around the world, now we are going to propose that you be the mayor of Betanzos.* Closing the circle of an incredible journey, Téllez agreed and moved back to his hometown, where he served as mayor for almost six years.

Yet in August 2020, Mayor Téllez was stunned to be visited by state security agents during the darkest days of the Áñez regime. They informed him that he was being charged with sedition and terrorism for reasons that apparently had to do with his participation in protests in the city of Potosí against the coup leaders. This was a time in which the coup government was persecuting hundreds of political opponents who had remained in Bolivia. When his family and colleagues in Halifax learned that he was being threatened with prison, they took a series of steps, including contacting the Canadian government and notifying high-profile experts on Bolivia, like Kathryn Ledebur, the director of the Andean Information Network, who provided updates about his situation to different media outlets.

Fortunately, before the government's persecution of Téllez got much further, the elections in October put an end to the dangerous charade. Things brightened considerably for him later when he was called to serve in public office again, this time as the head of Potosí Department's new lithium office. Making the long commute each day from Betanzos to Potosí, Téllez was a vigorous advocate for both the vision of the national lithium project and the potential of Potosí to play a supportive, yet essential, role in the project, especially in relation to battery development and production.

Téllez was responsible for assembling a small team of people for the lithium office who had different kinds of expertise: geology, environmental engineering, economics, and administrative law. He considered the legal and regulatory dimensions of lithium as important as the geological and chemical. As he explained, he wanted to make sure that any future national lithium legislation recognized, as he put it, the fact that the "battery transformation plant . . . is in our territory, so it would be very serious not to have anything to say or anything to do regarding such an important resource." One of Téllez's first moves was to make the suggestion—not a demand—to YLB that it "decentralize" its management structure and divide its headquarters between its offices in La Paz and a new administrative base in Potosí. Although Téllez counted on support for this proposal from the president of YLB, who was himself a potosino (Carlos Ramos, see chapter 3), the idea was never seriously considered on the nineteenth floor of Edificio Hansa.

Téllez and the rest of the small team in the lithium office realized early on that they would have very little to contribute to the relatively advanced process of lithium industrialization taking place in the far southwestern corner of the department, despite their enthusiasm for the project and a clear willingness to offer their support to the national government as a political olive branch—"Energy that Unifies." But the potential role of both the city and department of Potosí in YLB's battery operations was a very different matter. On the one hand, according to Téllez, the city was in the best position to provide a range of assistance, including researchers and students from the university, building contractors, food services, transportation, and other logistics for the growing facilities.

But on the other hand, Téllez wanted to find ways for the department to participate in the development of lithium-ion batteries that went beyond the laboratories, buildings, and workers that were already in place. In particular, his team envisioned a new extractive industry, perhaps directed by the department itself in collaboration with—but independent from—YLB, that would provide other elements for batteries apart from lithium. As he explained,

> We understand that the central state is the one in control, but we believe that we [as a department] have a unique role to play in the production of batteries. Batteries need . . . other elements, metallic elements. And . . . several of these metals exist in our territory.

What other elements? Cobalt? Nickel?

> Yes, and what we are hoping is that the government of Potosí will be prioritized for the investment in the search for these minerals. The [national] government is focused on the exploitation of lithium, we understand that, but we also believe there will be an opportunity to create the supply of these other minerals that are used in lithium batteries. If these other metals are found in Potosí, in our territory, there is no reason why they should have to be imported.[7]

Hitoshs Álvarez Ríos, a member of the lithium office and an environmental engineer, put it this way: "We are trying to figure out how we can revolve around the lithium productive chain without becoming too far detached from it, if that makes sense."

Despite this expansive vision—the hope that Potosí would someday find itself at the center of a separate raw materials industry, one in which it would both enjoy a "privileged position" (as Téllez imagined) and reap the material benefits—the department's various proposals for greater influence were mostly ignored by YLB officials in La Paz.

Although the posture of Potosí toward the national lithium project had shifted radically in only a few years, its aspirations for greater relevance were thwarted in part by the legacy of its historic opposition. Potosí's well-established record—stretching back into the early 1990s (see chapter 1)—of politicizing lithium as a largely failed strategy of coercive diplomacy cast a shadow over its relatively sudden course reversal. Nevertheless, the city of Potosí *would* have a more limited role to play in the life of YLB's battery operations, if only due to its coincidental proximity.

FIGURE 11. Rusting ruins of a former COMIBOL tin smelter, La Palca, 2022. Photo by author.

A PLACE WITHOUT LIMITS IN A COMIBOL GHOST TOWN

Giovana Díaz Ávila had been marked for a life of scientific distinction while still an undergraduate at the *Universidad Mayor de San Andrés* (UMSA) in La Paz. In July 2006, the city of Oruro hosted the 7th National Congress of Metallurgy and Materials Science. The event was held in the wind-swept altiplano city, famous for its mining history and UNESCO-listed carnival, in part to celebrate the 100th anniversary of the *Facultad Nacional de Ingeniería* (National Faculty of Engineering), which is part of the *Universidad Técnica de Oruro* (UTO). Among the dozens of papers presented at the congress, one was given on the "effects of mining waste on the durability of concrete" by Juan Carlos Montenegro, who would become the first head of YLB more than a decade later. But in order to highlight the importance of this particular congress, the event committee also organized a contest for young researchers, based on the results of studies conducted as undergraduate students. Díaz's prize-winning project was based on a study of the "Implementation and adaptation of the AFSolid 3.3 solidification simulator to the sizing of foundry feeding systems."[8]

A few years after graduation, Díaz received another accolade, this time one that was even more transformative for both her emerging career and Bolivia's emerging lithium project. Soon after launching its lithium master plan and breaking ground for the first buildings at the Salar de Uyuni, the Morales government realized that one of the main impediments to its lithium strategy was the dearth of Bolivian scientists who had either theoretical or practical expertise in many of the procedures that would be necessary for the project, from brine evaporation to the production of lithium-ion batteries.

To address this problem, the Morales administration decided to negotiate with governments from countries that *did* possess this expertise. The idea was to send cohorts of promising young Bolivian researchers abroad for extended periods of time, where they would receive advanced training and form international scientific networks. All of their expenses would be covered by these Bolivian government grants, but the candidates had to agree to return to Bolivia at the end of their training and put their new knowledge and skills to work in building the state lithium project from the ground up.

In 2010, the first two Bolivians were selected for this program, including the twenty-eight-year-old Giovana Díaz.[9] She was sent to Japan, where she joined the research group of Professor Kazuyuki Hirao of Kyoto University, which was conducting studies of ceramics, new energy technologies, and nanotechnologies. Díaz spoke with reverence about "Sensei Hirao," who was her scientific mentor during the nearly two years she stayed in Japan, a life-changing experience that opened her eyes to the world of international research, the use of advanced technology like electron microscopes, and the chance to gain practical experience with lithium-ion battery production in collaboration with Japanese companies.

I asked her if she was able to meet with Akira Yoshino during her stay in Japan, the researcher at the Asahi Kasei chemical company who would later (in 2019) be awarded the Nobel Prize in Chemistry along with John Goodenough and M. Stanley Whittingham for the "development" (not invention) of the lithium-ion battery. Although she didn't meet Yoshino, she was able to make regular visits to research centers across Japan with various connections to battery development. One of her visits was to the University of Kitakyushu, where a research group was working on an experimental method of lithium extraction that used a "green solvent" applied to brine that the Japanese researchers had obtained from the Salar de Uyuni.[10]

However, in discussing in detail the ways in which her time in Japan had changed her life, both scientific and personal, Díaz drew a surprising contrast. From her perspective, the key differences between the scientific cultures of Bolivia and Japan were technological and institutional, rather than pedagogical.

> When I graduated from university [in Bolivia], I felt like I had been well educated. And after arriving in Japan, I saw that we were at the same level [of knowledge]... The difference is simply a matter of technology, a matter of order, of discipline, of ease of doing research. In the university [in Kyoto], what amazed us is that ease, there is very little bureaucracy, both in the public part and in the more private part, because we were also affiliated with a private research center ... Because it was private, it was very efficient. If you wanted to do a research project, you could do it in a week, you had everything at your disposal, all of the equipment and facilities of the university.

More personally, Díaz's stay in Japan transformed her into an international researcher while reaffirming her commitment to pursue a scientific career that both advanced knowledge and supported the public good. As she told me, her only regret was that more young Bolivian scientists didn't have the same opportunity to travel abroad to deepen their training and build research networks.

Soon after returning to Bolivia in 2012, the now-thirty-year-old Díaz went to work right away for the Gerencia Nacional de Recursos Evaporíticos (GNRE), the specialized agency within the state mining company (COMIBOL) charged with overseeing the incipient lithium project, which now included lithium-ion batteries (YLB, the state lithium company, wouldn't be established until 2017; see chapter 2).

Díaz was a member of a small team that was asked to draw up a detailed technical proposal for the battery facilities, in large part modelled on what she had learned and seen during her time in Japan. As she explained, the team's remit from the Morales government was wide-open:

> They told us, do not limit yourselves based on the experiences we have had, we will do whatever needs to be done. We need a center and we need pilot

projects . . . Because in Japan itself, what caught our attention is that after the war, the world war, they rebuilt . . . by having more than one hundred research centers throughout the country . . . And we thought: Why not do the same thing here in Bolivia, except only for the lithium industry? The idea was not just to train simple technicians, people to whom you'd say, "you have to do this, you have to follow this recipe." No, we believed that we had to go further and for that we needed a place where we had all the conditions to do it, where the project could develop, where we could do research, and where we could do the analyses we needed to do without limits.

But while Díaz and the small team of Bolivian scientists worked on the proposal for the country's battery operations, an immediate problem arose: where was this place without limits to be located? Here, apparently, there *were* limits, limits defined by age-old considerations of patronage and a kind of institutional *caudillismo*, whereby the personal preferences of the boss determine major decisions. When Díaz's team had been asked about a location very early on in the process, they recommended a site in El Alto, the rapidly growing city on the altiplano above La Paz. They reasoned that this would put the battery center near a number of universities, including UMSA (Díaz's alma mater), and on the doorstep of one of Bolivia's two major international airports, which would have a range of advantages, from supply logistics to access for visitors. Yet this initial recommendation was never seriously considered.

At the time of these deliberations, the Morales government had just appointed a new mining minister, Mario Virreira Iporre. Virreira had spent his youth in the COMIBOL mining camp at Colquiri in La Paz Department, before moving to the city of Potosí for his higher education. He graduated from UATF with a degree in civil engineering and then worked for COMIBOL before entering academia, where he taught at UATF and served in academic administration, eventually becoming rector of the university. In 2005, in the historic wave that swept MAS into power, he was elected "prefect" of Potosí Department as a MAS candidate (the post was later converted into a governorship). In 2012, he received the cabinet-level post of mining minister as part of a reshuffle during Morales's second term.

However, many years before ascending to the heights of academia and political influence, Virreira had worked as a young engineer at an ill-fated COMIBOL installation, one whose dark history has largely been forgotten. In early 1971, during the short-lived (ten months) regime of the leftwing military leader Juan José Torres, the Bolivian government signed a wide-ranging bilateral trade agreement with the Soviet Union that included the extension of credit, which Bolivia was then supposed to use to purchase Soviet goods, technology, and services.

As part of this agreement, the Soviets agreed to provide technical assistance to build a series of tin smelters, including one about twenty kilometers from the city of Potosí, in a small valley near the community of La Palca. The Soviet Union was keenly interested in both gaining privileged access to Bolivian tin and in establishing a political foothold in a region that had traditionally been under the interventionist influence of the United States, a legacy that included the active US involvement in the capture and execution of Che Guevara in Bolivia in 1967.

The tin smelter project at La Palca, under the supervision of Bolivia's state mining company, was a multifaceted disaster from the beginning. When Torres was ousted in the bloody coup that brought the pro-US Hugo Banzer to power, the Soviet technicians were expelled in the early stages of construction; from then on, Bolivian workers took over the building process, "despite geological studies that cast doubts on the geological structure of the location."[11] As might have been predicted, a series of massive landslides repeatedly destroyed the emerging structures, causing the project to be delayed by over ten years. In 1982, the tin smelter was finally finished, but before it could be tested (using tin concentrates from Potosí's Cerro Rico), "a portion of the plant caught fire, apparently because of faulty diesel burners."[12]

Having repaired the fire damage over several additional months, COMIBOL was finally able to put the smelter into operation. However, within just days, *campesinos* from the small village of La Palca found themselves bombarded with heavy pollution from the plant, which was emitting 120 tons of sulfur every day, much of it in the form of sulfuric acid, which "did severe damage to plant and animal life in the region."[13] After a period of negotiations, COMIBOL reached a settlement with La Palca, which included payments to the community and an agreement to purchase all the land in a one-kilometer radius around the smelter in order to create a buffer zone.

To reduce the sulfur pollution, COMIBOL attempted a series of technical fixes, including the construction of a tall smokestack, which itself suffered damage when "the presence of fluorine and chlorine" in the tin led to corrosion.[14] And beyond all of the environmental and structural problems, the tin smelter was also a financial disaster for the state: in the end, the facility cost Bolivia over $80 million to build (about $600 million adjusting for inflation) even though it was initially budgeted to cost only $8 million.

The dark cloud over the tin smelter remained. Although finally operational, despite the many problems, it was shuttered within just a few years during the "shock doctrine" neoliberalization of the Bolivian economy by the government of Víctor Paz Estenssoro. As part of this rapid turn toward "structural readjustment" under what came to be known as the New Economic Policy, COMIBOL was closed down between 1986 and 1987 and its mining operations privatized, leaving tens of thousands of state mine workers—miners, administrators, engineers, and others—out of work. The tin smelter at La Palca was abandoned, which was no doubt welcomed by the neighboring community, but the industrial equipment, small administrative building, and all of the land that was part of the buffer zone remained Bolivian state property. Over the subsequent decades, the former tin smelter became a ghost town, its rusting structures enveloped by a dangerous pall of lingering toxicity.

And yet, it was precisely this unlikely location that was chosen by mining minister Virreira as the site of Bolivia's future battery research and assembly center, the same site where he had worked as a young COMIBOL engineer. Although the land and buildings at La Palca still belonged to COMIBOL, which meant the government didn't have to purchase any new property, the location otherwise met none of the criteria that Giovana Díaz and her team had proposed for the development of the country's lithium-ion battery industry.

About thirty minutes from the city of Potosí in a rural valley off the main Potosí—Oruro national highway, La Palca is nevertheless far enough from the city to make it relatively difficult to reach. And Potosí itself, despite its history, is notoriously isolated compared with most of the other departmental capitals in Bolivia. In fact, during its many mobilizations, COMCIPO, Potosí's militant civic movement, regularly pushed for the construction of a new airport as part of its demands on the national government. The old Capitán Nicolas Rojas airport receives few commercial flights and is often nonoperational for months or even years at a time. To reach Potosí from La Paz by bus requires a grueling journey across long stretches of the altiplano that can take up to ten hours.

Despite it all, in 2012, Bolivia quietly launched its battery operations at the abandoned tin smelter. The plan called for three different battery divisions at La Palca: a center for research and development; a laboratory for producing battery cathode materials; and a facility for assembling lithium-ion batteries at an experimental, or "pilot," level. Although the original lithium master plan centered on what it called "100% Bolivian" capacity across the entire process, from lithium extraction to the manufacture of batteries, it was understood from early on that this essentially political vision would be impossible to realize for the "downstream" stages, especially those required to produce high-capacity—or, really, *any* capacity—lithium-ion batteries. Indeed, as we have seen, this realization was the reason the Bolivian government had sent an initial group of young scientists abroad in 2010, a first step that was dramatically expanded in 2014 with the opening of the 100 Grants program, through which dozens of junior Bolivian researchers would go on to receive advanced training at institutes around the world.[15]

But until this long-term investment in "100% Bolivian" research and production capacity could bear fruit, the government needed a different approach. They turned, therefore, to a strategy that might be described as "50-50 Chinese-Bolivian," followed by "50-50 French-Bolivian." In May 2012, Bolivia signed a contract with Linyi Dake Trade Co., a subsidiary of the Chinese battery technology company Shandong Gelon Lib Co., to deliver a "key in hand" lithium-ion battery assembly facility. The contract called for the Chinese company to ship all the equipment necessary for the assembly of batteries from preexisting components; install the equipment at La Palca; and train GNRE/COMIBOL employees in the use of the equipment. And where was this "pilot" plant to be installed? In an old building that formerly housed administrative offices of the long-abandoned tin smelter. Linyi Dake was required to refurbish the existing building so that the rooms could be repurposed into a small-scale battery factory.

While this refurbishment was taking place, the Bolivian government moved forward with the second pillar of the project at La Palca: the installation of another pilot plant, this one for the much more complex production of cathode materials, which are the chemical heart of lithium-ion batteries. In 2015, another "key in hand" deal was signed, this time with a relatively unknown French engineering company called Greentech, based in Grenoble, which specializes in the manufacture of high-end solar panels. Although they have a side business delivering so-called turnkey projects, where they install equipment and provide the knowledge necessary for their partners to use it, they had no prior experience with the lithium-ion-battery industry.

Nevertheless, Greentech and the Bolivian government signed a $4 million contract for the French company to set up two separate cathode synthesis "lines," one leading to LMO (Lithium Manganese Oxide), the other to NMC (Lithium Nickel Manganese Cobalt Oxide). The necessary labs, some of which required sterile conditions and controlled climate, also had to be installed in a refurbished section of the old tin smelter.

Global lithium analysts like Lukasz Bednarski were either befuddled or dismissive of the fact that Bolivia chose a "PV equipment manufacturer" to jumpstart the most ambitious part of its battery strategy. As he puts it, "The Bolivian state could have easily worked with the crème de la crème of the lithium economy. For unknown reasons it chose not to."[16] But this critique from the bird's-eye perspective of global value chains and market dynamics has little to say about how the cathode materials—or, for that matter, the battery assembly—projects actually developed at La Palca. For example, Greentech organized a series of visits by Bolivian battery scientists and technicians to the Grenoble branch of the French Alternative Energies and Atomic Energy Commission, or CEA, where they received hands-on training in cathode synthesis. Following this, Greentech oversaw testing and implementation at La Palca for almost six continuous months leading up to the inauguration of the cathode materials facility in 2017.[17]

Finally, the third pillar of Bolivia's battery operations at La Palca required the only major construction of a new facility. Although a number of dedicated offices for research and development had also been installed in the refurbished tin smelter buildings, the plan was to design and build what would be the most advanced battery research center in Latin America—again, in the most unlikely of locations. In 2018, the new Bolivian state lithium company YLB (established in 2017) signed a contract with the Potosí construction company Quintanilla & Quintanilla to build a forty-thousand-square-foot facility, a project that was supposed to take two years to complete.

The design of the laboratories and sourcing of advanced technology, including a high-resolution transmission electron microscope (TEM), which allows for crystallographic imaging at an atomic scale, was under the responsibility of Solar Datalab, a Colombian firm that specializes in "analytical instrumentation" and laboratory installation. After a delay of almost a year due mostly to the intertwined crises of the coup and Covid-19 pandemic, the stunning new research center—known officially as the Centro de Investigación en Ciencia y Tecnología de Materiales y Recursos Evaporíticos de Bolivia—was inaugurated in December 2021.

And what about Giovana Díaz, the pioneering Bolivian battery scientist? After playing a key role in the development of the facilities at La Palca, including the new research center, which was her main responsibility and passion, she was elevated to director of all battery operations, which required her to move to YLB headquarters in La Paz. Although this was a major promotion, it took her away from the advanced laboratory work that she had valued most since her days at Kyoto University. Despite the historic fact that a woman was in charge of the most important technology project in Bolivia, Díaz did not thrive in her new administrative role, and she was let go by YLB after only a few months.

Although this was a great professional blow to her, especially given the fact that she had given her entire career to Bolivia's state lithium project, it was perhaps a blessing in disguise. After a period of uncertainty, she managed to secure a position as a research professor in the Instituto de Investigaciones en Metalurgia y Materiales, an academic unit in her alma mater UMSA that was directed by Juan Carlos Montenegro, her old boss and ex-president of YLB. And in 2023, Díaz—along with two others—was awarded the first patent in Bolivia for a lithium technology, a process that uses an acid to leach lithium ions from "commercial" grade lithium carbonate as a way of purifying it further into "battery" grade lithium carbonate (that is, with a purity higher than 99.5%).

GETTING TO $LiNiMnCoO_2$

In a remarkable coincidence, I began conducting research on Bolivia's lithium project in the same year that the Nobel Prize in Chemistry was awarded to three scientists for the development of the lithium-ion battery. Why these three? The answer helps explain the basic structure of a lithium-ion battery. M. Stanley Whittingham's research in the early 1970s, part of an experimental project to develop alternative energy sources during the oil crisis, led to the creation of a cathode, or positive electrode, that could absorb and store lithium ions. Somewhat later, John Goodenough proposed a different cathode chemistry, one that had much greater energy density. In the mid-1980s, Akira Yoshino solved the problem of the anode, or negative electrode, in the battery. In his early battery prototype, Whittingham had used an anode that was highly reactive, causing the battery to explode, which is one of the reasons his research did not lead to the production of commercially viable batteries.

But Yoshino found that a carbon-based anode can intercalate, or store, lithium ions and remain stable. With the basic design for the positive and negative electrodes in place, the lithium-ion battery revolution could begin.

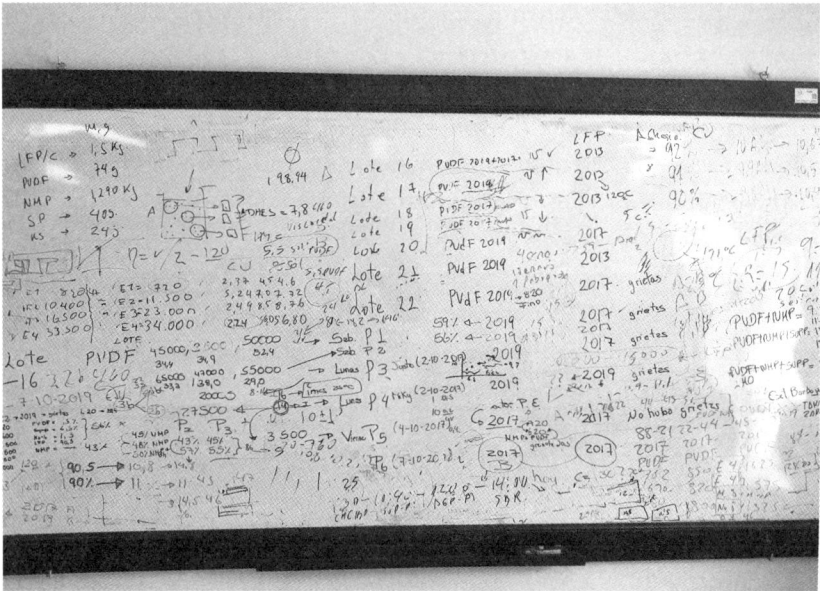

FIGURE 12. Calculations and figures, YLB battery facility, La Palca, 2022. Photo by author.

As the press release for the 2019 Nobel Prize in Chemistry explains, "The advantage of lithium-ion batteries is that they are not based upon chemical reactions that break down the electrodes, but upon lithium ions flowing back and forth between the anode and the cathode."[18]

These ions flow back and forth through the cycle of charging and discharging through a liquid medium, the electrolyte, which consists of lithium salts mixed with different kinds of organic solvents. Electrical current, and thus power, is generated when lithium ions move from the anode to the cathode during discharge. When the battery is then in its charging cycle, the lithium ions return to the anode.

Although the precise composition of all three of these materials—the cathode, the anode, and the electrolyte—determine the specific profile of a battery, it is the cathode that receives most of the attention. As the positive electrode, cathode composition is much more complex from the perspective of electrochemistry and chemical engineering. Even if researchers continue to work on new designs for the anode and electrolyte, it is with the cathode that these innovations have made the greatest difference.[19] Based on current technology, there is no perfect or ideal battery, even for specific applications, like the lithium-ion batteries used for EVs. Leaving the anode and electrolyte aside, all

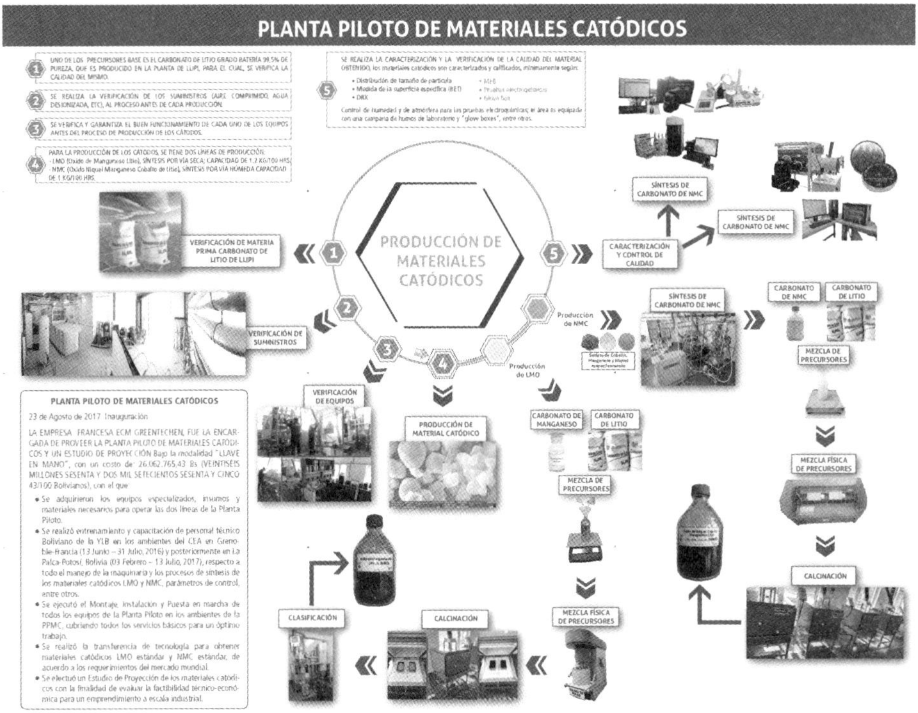

FIGURE 13. Description of cathode materials production from YLB's 2018 annual report.

cathode chemistries involve trade-offs with respect to a number of key characteristics: energy density, safety, performance, cost of materials, and lifespan.

In the YLB cathode materials laboratories at La Palca, Bolivian technicians work on two distinct "lines," as they are called. Although the process is similar, each cathode chemistry requires its own series of steps to arrive at the final compound, which can then be used in the assembly of a battery (see Figure 13). The two lines chosen by YLB (then GRNE/COMIBOL) were selected for specific reasons. The LMO (Lithium Manganese Oxide) cathode is a well-established chemistry that allows for fast charging but has a shorter lifespan and lower energy capacity than others. LMO cathodes are used in batteries for power tools and other smaller electronic devices, but not EVs. LMO cathodes are also much cheaper than other chemistries because manganese is relatively abundant.

The second line being developed at La Palca leads to NMC (Lithium Nickel Manganese Cobalt Oxide, or $LiNiMnCoO_2$), which is a very different cathode than the LMO. NMC also contains manganese, but it is the inclusion of nickel and cobalt that has made NMC cathodes the most widely used in high-capacity batteries, most importantly in the batteries that are assembled into the large battery banks that power EVs. The reason for this has to do with the specific characteristics of NMC cathodes, which have both a high energy capacity and a long cycle life. For example, batteries with LMO cathodes can produce 100–150 watt-hours per kilogram (Wh/kg) through 300–700 discharge/charging cycles, while batteries with NMC cathodes can produce 150–220 Wh/kg through 1000–2000 cycles, depending on a range of things like "depth" of discharge and ambient temperatures—think Minneapolis vs. Houston.

However, the price of both nickel and cobalt makes NMC-based lithium-ion batteries much more expensive to source and produce. Furthermore, nickel and cobalt mining take place under devastating environmental and social conditions. Most of the cobalt in the world is mined in the Democratic Republic of Congo, where Chinese companies dominate the industry. Miners—including children—work under atrocious circumstances to extract a metal whose scarcity, price, and association with underlying sociopolitical violence qualify cobalt as an emerging conflict mineral. But because of the ubiquity of NMC cathodes in the global market for EVs, YLB decided that it must put an NMC production line at the center of its battery strategy.

Within the refurbished rooms of the old tin smelter at La Palca, YLB's battery researchers and technicians don't think about these wider political economic dilemmas or the reasons why the Bolivian government chose to produce certain kinds of cathodes and not others. Instead, they focus on the different steps in the painstaking and technologically fraught process that culminates, in the best of circumstances, in a surprisingly small amount of compound that can be "classified" as either LMO or NMC. These compounds are kept in small bottles whose modest quantities testify to the essentially experimental nature of the process.

Leaving aside the steps in this process that are more mechanical, like mixing and calcination, which requires the materials to be heated to high temperatures, the YLB technicians at La Palca are most challenged by the different phases of characterization, quality control, and classification. These take place through electrochemical testing, including through the use of "glove boxes," and most importantly through the use of a scanning electron microscope, or SEM. In a small office, a technician who I'll call Jorge Carlos allowed me to observe him as he analyzed cathode materials from the LMO production line.

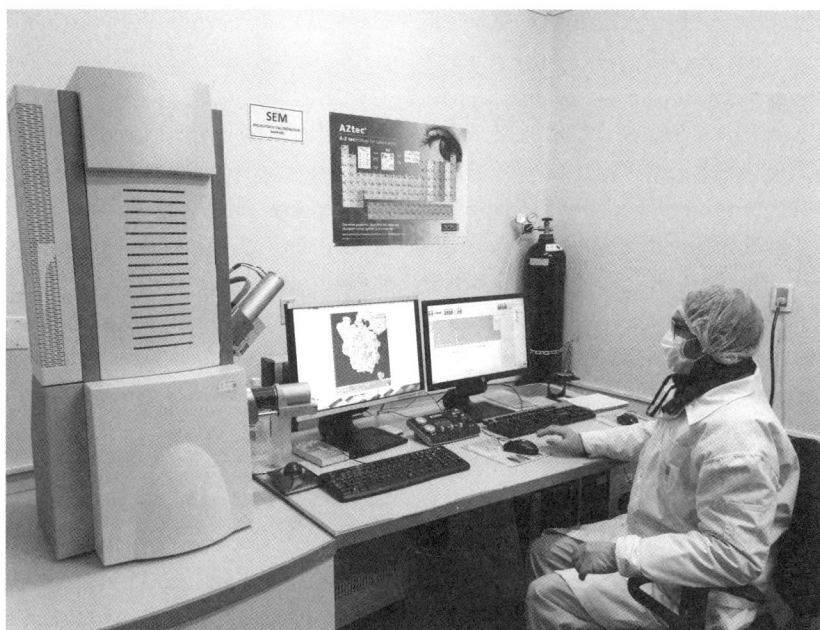

FIGURE 14. YLB technician analyzes crystal structure of cathode sample, La Palca, 2022. Photo by author.

It was a quiet but extraordinary scene: in an old, if "refurbished," building at an abandoned COMIBOL tin smelter in the Bolivian *campo*, a highly trained technician was using a $75,000 electron microscope to beam electrons onto a small sample of cathode material. He then used equally sophisticated software to analyze the resulting crystal structure based on carefully defined criteria around particle size, particle distribution, and the extent of something called "dynamic recrystallization." I asked Jorge Carlos how he conducted the classification. He said that he relied on both a visual interpretation of the crystal structure—which was displayed on the large screen at two thousand times magnification—and an evaluation of the accompanying analysis program. Given that he was verifying that the sample was, in fact, cathode-ready LMO, he was quick to point out the predominance of both manganese and oxygen.

And beyond the electron microscopy, what also makes Bolivia's efforts to develop the homegrown capacity to produce battery cathodes so astonishing is the way in which its technical infrastructure is enmeshed in multiple worlds of scientific nationalism, each with its own deep history. This narrative of infrastructural enravelment could be seen everywhere at La Palca, including at Jorge Carlos's workstation. The SEM had been built by a storied Czech technology company with its origins in socialist Czechoslovakia's small postwar electron microscope industry, centered in the Moravian city of Brno. The CEO of TESCAM, which made the Vega 3 SEM used by YLB ("high performance at an attractive cost"), got his start in the 1970s as a young engineer working for a microscope company called, yes, Tesla.

And as Jorge Carlos explained, he combined an interpretation of the crystal structures of different elements in the cathode sample with the use of a "materials characterization" software package. For this, YLB used a platform developed by Oxford Instruments, one of the first spin-off companies associated with the University of Oxford. Founded in 1959 by Martin Wood, an engineer and business developer based in the university's Clarendon Laboratory, the company built the first commercial magnetic resonance imagining (MRI) scanner. Oxford Instruments (its trademarked motto: "The Business of Science") sold YLB its "AZtec" package for the SEM, a curious name that is derived from the first part of the tagline "A-Z technology for nanoanalysis." To go along with its software, Oxford Instruments had supplied a handy poster of the periodic table, which Jorge Carlos had affixed to the wall above his desk.

Jorge Carlos and his colleagues, working on their own distinct steps in the process of getting to NMC, or LMO, did their best to apply their training while at the same time following what another poster in a common hallway called the "Ten Commandments of Public Service." In the meantime, a very different process was taking shape in yet another former building of the abandoned tin smelter: the actual fabrication of lithium-ion batteries.[20]

PILOT FORDISM AND SOCIAL PRODUCTION

In 2016, a couple years after the initial launch of operations at La Palca, workers found a strange contraption tucked into the corner of one of the buildings of the old tin smelter that had *not* been refurbished. Although the Chinese company Linyi Dake had done what it could to convert decades-old abandoned structures into places where advanced research could be conducted and batteries assembled under less-than-optimal circumstances, the larger "industrial complex" at La Palca was filled with a wide assortment of industrial detritus. Even when the site became fully operational years later, part of the grounds still looked something like a junkyard.

But workers discovered something very interesting among the scrap: a four-wheeled transport cart that seemed to have been powered at one time by batteries. In examining this odd vehicle further, they saw that it had been manufactured in the Soviet Union and was therefore one of the pieces of equipment that had managed to survive as a reminder of the ill-fated origins of La Palca itself. The vehicle's red paint had faded and its battery panel had been pulled off, exposing its internal cables. Its two control levers were falling apart and the wheels were broken. To round out the playful absurdity of the scene, an unknown GNRE worker had christened the rusting hulk by writing its new name on the side: "El Chapulín." This funny-looking red vehicle was now named after El Chapulín Colorado, The Red Grasshopper, the beloved and bumbling superhero made famous throughout Latin America by the Mexican comedian Roberto Bolaños in his TV show from the 1970s ("More agile than a turtle, stronger than a mouse, nobler than a head of lettuce . . . ").

Yet in a spirit that was noticeably distinct from the one prevailing in either the nearby research center or, somewhat later, the cathode materials laboratories, the workers in the pilot battery facility decided to bring El Chapulín back to life. Because they were learning how to assemble lithium-ion batteries from scratch, and were already building a line of 10 Ah battery cells, they decided to construct their first battery *pack* for their improbable first EV. Using an approach that, with some modifications, forms the base of all EVs, they wired together six battery blocks, each with ten battery cells, creating an integrated power source for El Chapulín of a very modest 32 volts and 60 Ah. They then repaired the control levers, internal wiring, and wheels and gave their first EV a fresh coat of red paint. After more than four decades gathering dust, El Chapulín was on the move again, although its use was limited to the small, paved area just outside the battery facility.

As the first of the three divisions at La Palca, the Planta Piloto de Baterías (PPB) was marked by this culture of bricolage from the beginning. Although the Bolivian government had bold ambitions for the production of lithium-ion batteries, even more than for either battery research or the production of cathode materials, it was with battery production that the gap between the state's vision for the wider lithium project and its realization in practice was the widest. From the outset, this vision imagined a revolutionary new "productive chain," in which battery grade lithium carbonate was produced at an industrial scale in Bolivia and then used in the industrial production of high-capacity lithium-ion batteries, also in Bolivia. What does the "industrial production" of lithium-ion batteries mean in this sense? By way of illustration, CATL, the Chinese company that is the world's largest battery supplier, has a factory in southwestern Guizhou province that can produce about thirty-one million batteries per year, or about eighty-five thousand *per day.*

In other words, Bolivia might very well develop the capacity to produce lithium carbonate at an industrial level, but its projections for batteries must be understood quite differently. The imagined completion of the lithium productive chain was always a political economic gesture with symbolic value, rather than an actual plan that would lead to the emergence of a Bolivian battery industry. But this doesn't mean the pilot battery factory was a failure; quite the opposite. Understood through its symbolic and social valences rather than through its commercial outcomes, the pilot battery operation at La Palca becomes an example of what might be thought of as *divergent* ingenuity, of production beyond—or even against—the boundaries of state ideology.

This divergence took shape in a paradoxical way. On the one hand, the technicians at La Palca took the basic "key in hand" set-up installed by Linyi Dake and used it to experiment with assembling lithium-ion batteries through different processes of trial-and-error, including using a reverse engineered 800-milliampere Nokia cell phone battery to calibrate the assembly line. They also followed a flow chart provided by Linyi Dake, which showed how to complete the assembly line, from the insertion of the cathode and anode materials, to the sealing of the battery cell itself. Although the battery workers, including the managers, at La Palca were always circumspect about the provenance of the different materials that were used to assemble batteries, most of the key ingredients—cathode, anode, electrolyte—were sourced from Chinese suppliers. Even if a handful of batteries *were* eventually made with cathode material produced at La Palca, the vast majority of the Bolivian cells were based on LFP (lithium iron phosphate) chemistry, which, as we have seen, was not one of the two local cathode production lines.

But as Giovana Díaz was quick to point out, although the materials for the cathodes, anodes, and electrolytes were sourced commercially, YLB "NEVER" (her vehemence suggests all-caps) imported prefabricated battery cells for the different devices that were produced in the PPB. Instead, the overriding consideration was to develop the assembly line itself as a technology, even if, in the end, it was never put to use in *industrial* production. What did it mean to be able to use a conventional Fordist approach to assemble a lithium-ion battery? How could each step in the assembly process be both segmented as an area of specialization and integrated at the same time? Questions like these shaped everyday labor in the PPB, as well as local decision-making, all of which fit well with the national lithium master plan and its symbolic gesture toward a future battery industry in Bolivia.

On the other hand, this "pilot Fordism" was never actually used to manufacture batteries for the market—*any* market, especially the market for EVs. This wasn't just an effect of the growing realization over time that the attempt to develop the capacity to produce the kinds of battery cells and battery packs used in commercial EVs made little sense. This was true even though YLB did sign an agreement to produce a small number of battery packs for Quantum Motors, a Bolivian EV start-up (see chapter 5). Rather, it was more that managers and technicians at La Palca, far from the nineteenth floor of Edificio Hansa in La Paz, wanted to produce things of real social value, even if they played no role in the global energy transition, and even if the devices themselves contained not a trace of Bolivian lithium.

Perhaps the most important example of this alternative social production in La Palca involved a project in which the PPB assembled batteries that were used for household electrification in rural villages in the area. Using components purchased from the German company Zimpertec, which specializes in "off-grid" electrification systems that "contribute to a flourishing social life and create equal opportunities for everyone," PPB technicians began installing these battery-powered solar systems in communities, where they were used to provide lighting in what are often one-room adobe dwellings.

Although this project had nothing to do with Bolivia's lithium master plan, it was such a success with local communities that the national government eventually incorporated it into a preexisting initiative that had been launched in 2008, the Electricity Program for Living with Dignity, which was originally based on the use of hydroelectric power. However, even though the embrace of the rural solar systems—which used lithium-ion batteries assembled at La Palca—allowed the hydrocarbons and energy ministry to emphasize its commitment to green energy, they were never distributed on a national scale.

But the value of the rural electrification project was clear to the mangers and technicians at La Palca. It was something that gave meaning to their work in ways that were tangible, visible, and consistent with a wider vision of public service, again, even if this meant diverging widely from the underlying rationale for the revivification of La Palca as the implausible center for the country's future battery industry. Even more, it was through these forms of

FIGURE 15. Lithium-ion batteries assembled for rural solar systems, La Palca, 2022. Photo by author.

nonindustrial production at La Palca that workers were able to reinforce their own affective ties with Bolivia's imagined futures, whether these involved lithium or not.

I asked José Erquicia Cruz to reflect on the shadow social production that was taking place at La Palca and its uncertain relation to the country's wider lithium master plan. Although he was only thirty-seven years old when I met him in 2022, he was one of the old hands at La Palca, having worked at the site with the small group of Bolivians who arrived in late 2013, even before the Chinese had finished with the refurbishment. He had grown up in the rough and tumble town of Villazón in the far south of Potosí Department on the border with Argentina, a place where goods of all kinds move back and forth under various shades of legality and danger. Through a combination of savvy and luck, he had managed to survive the regular culling and replacement of personnel, including the purges of YLB during the 2019 coup and its aftermath. He was hardened by almost a decade of work at La Palca, but also proud of what he and his colleagues had accomplished.

> *For us, for me personally, being here and being part of this work team is historic, because, well, as a chemist, I never thought I would be able to produce a battery in my country . . . From batteries to the other types of products we make here, it is something historic for us, because we are contributing as Bolivians, contributing to our country, and producing many things that we had never even thought of as chemists, or electricians, or engineers. It has been a great, great experience, and also an opportunity, even though what we are actually contributing here is just a grain of sand.*

CONCLUSION: LITHIUM DEBT AND THE LIMITS OF SYMBOLIC VALUE

By early 2023, the destiny of La Palca and Bolivia's battery aspirations had become clearer. As the hydrocarbonization of the wider lithium project accelerated, as commercial production above all else in the shortest amount of time became the new marching orders, policy-makers in La Paz began to look at La Palca as an experiment whose days were numbered. At a moment in which a historic reckoning was taking place in relation to the seemingly never-ending delays on the lithium industrialization side of the ledger, the battery strategy, which at least officially still pointed toward industrialization, made an easy target. When YLB policymakers began to dwell more and more on the amount of debt incurred over almost fifteen years of bottomless state investment in the lithium project, it was only a matter of time before the place without limits, as Giovana Díaz had imagined it, would face a reckoning.

Carlos Ramos, the head of YLB, tried to cast what was coming for the workers at La Palca, including the intrepid José Erquicia, in the best possible light. As he put it, the fate of La Palca was a direct result of what its personnel had managed to achieve. *They have already shown that it is possible, that we can produce cathodes, that we can produce batteries here in Bolivia.* And then he described the battery operation and its workers in a way that might, or might not, have intentionally drawn on a metaphor for the batteries themselves: *In a sense, they have already completed their lifecycle.*

FIGURE 16. YLB's new battery research center, La Palca, 2022. Photo by author.

But despite the unmistakable cloud hanging over La Palca and its otherwise deeply meaningful grains of sand, a major question remained: What to do with the gleaming forty-thousand-square-foot research center, which was still in the process of being outfitted with high-end equipment and advanced laboratories? Amid this broader uncertainty, the inauguration ceremony for the research center was nevertheless attended by the full suite of officialdom, including President Arce, Franklin Molina (minister of hydrocarbons and energy), Ramos, departmental officials, and YLB personnel from La Palca's different divisions.

Festooned with brightly colored floral garlands and wearing YLB hardhats, Arce, Molina, and Ramos were given a tour of the new building, which included presentations to the research center's star attractions: the various pieces of equipment that, taken together, were meant to secure

Bolivia's status as a future Latin American hub for advanced research not just on lithium but on the material foundations of energy more generally.

After peering at a nuclear magnetic resonance spectrometer, a massive X-ray diffractometer, and an inductively coupled plasma-optical emission spectrometer, among other pieces of high-end technology, the group finished the tour at the research center's showstopper: a towering, brass-colored, high-resolution transmission electron microscope, or TEM-HR. Although YLB officials were always reticent to talk about how much their TEM-HR cost, the price tag on new high-resolution TEMs can approach $1 million. Compared with the older Czech-built SEM in use at La Palca, this brass monolith gives researchers the ability to analyze the atomic structures of particles at magnifications up to fifty million.

Although the YLB team didn't give a working demonstration of their prized possession, Arce seemed genuinely impressed. In an enthusiastic speech at the event, he gave no indication that the new research center at La Palca—whose building design includes systems that protect against landslides and runaway electromagnetic fields—was under threat because of the shift of orientation in the wider national lithium project.

But if the research center will be preserved, the same almost certainly cannot be said for the other divisions at La Palca, which were originally conceived as the final links in Bolivia's length of the lithium "productive chain." Instead, Bolivia's stretch of chain will likely end with the industrial production of lithium carbonate at the Salar de Uyuni. Bolivia's quixotic foray into the worlds of lithium-ion batteries, from cathode synthesis to battery assembly, will thus take its place as yet one more moment in the country's long, and still unfolding, struggle with the many sides of lithium. And what about the old, if refurbished, buildings of the former state tin smelter? Even if the research center will live on as a monument to Bolivia's growing scientific capabilities, the other structures at La Palca will be abandoned once more, left to gather dust, waiting to be reclaimed yet again.

_____

FIVE

Electric Ambitions, Made in Cocha

IN 2019, A CURIOUS VEHICLE was unveiled by a small group of engineers and technicians at Bolivia's experimental battery facility in the village of La Palca on the outskirts of the city of Potosí. It looked like a small dune buggy with four wheels, a stainless-steel roll bar without doors, and a small grey metal hood painted with the company logo of Yacimientos de Litio Bolivianos, or YLB. The little car weighed about 620 pounds and was christened "Kachi Car Electric," using the Quechua word for "salt." The Kachi Car was, in fact, the first EV built by the Bolivian state lithium company. Using battery packs assembled in the pilot plant at La Palca, which incorporated battery cells based on LFP chemistry (with a capacity of sixty-four volts and one-hundred amp-hours), the little car managed to drive about thirty miles on a single charge. The YLB technicians even managed to construct a charger for the Kachi Car, which had the capacity to recharge the vehicle in under forty minutes.

The Kachi Car had originally been designed with a very specific purpose in mind: to take part in an innovative competition in Bolivia launched by one of the country's most visible green and alternative energy NGOs, Energética, based in the city of Cochabamba, which, as will be seen below, has become a center of Bolivia's own version of the green energy transition. Energética was founded in 1993 by a couple of local energy researchers and entrepreneurs, Miguel Hernán Fernández Fuentes and Nepthalí Sierraalta Uría. Since that time, Energética has amassed a large portfolio of projects under the umbrella of what it describes as "energy for development," which seeks to make renewable energy and energy equality the bases for social and economic development.

In 2016, Energética announced that it was sponsoring a new road race for electric vehicles, the "Bolivian Solar Grand Prix," which would be open to

FIGURE 17. "Yes, I Am Electric," Cochabamba, 2022. Photo by author.

entrants that met a series of stringent conditions, including that each race team had built its own EV. For the first editions of the race, the winners in the different categories had been EVs designed by university teams. By 2019, however, the wider discourse around lithium and Bolivia's intention to complete the entire "productive chain" had reached a high point, just around the time YLB signed the ill-fated deal with the German company ACI-Systems (see chapter 2).

Although it was usually mentioned more as a tantalizing afterthought by different leaders, including Evo Morales, the production of EVs was invoked as the ultimate goal for a process of productive sovereignty in which Bolivian cars would be powered by Bolivian batteries made with Bolivian lithium. The design of the Kachi Car, therefore, was viewed by the small YLB crew at La

Palca as the necessary first step in realizing this ambitious dream on behalf of the country. The appearance of the Kachi Car in the Bolivian Solar Grand Prix would be an historic moment, a signal to the world that the Bolivian state had moved that much closer to its plans for true vertical integration, from the production of lithium carbonate to the manufacture of EVs.

However, like so much else in the longer saga of lithium in Bolivia, the destiny of the Kachi Car would be quite different than expected. Because the earlier Bolivian Solar Grands Prix had all taken place at the end of November, the planning for the 2019 edition happened to coincide with the aftermath of the national election, which led to a coup d'état and a year of chaos and violence. Under these circumstances, the solar-powered road race for 2019 was understandably canceled. The 2020 Bolivian Solar Grand Prix was also canceled by Energética as the country—like the rest of the world—did its best to cope with the catastrophe of the Covid-19 pandemic, which unfolded in the midst of national political turmoil, rampant corruption by the coup leaders, and a complete botching of the public health crisis by the "interim" government.

By the time Energética got around to organizing the fourth Bolivian Solar Grand Prix in 2021 after a two-year gap, much had changed—for Bolivia, for its lithium "master plan," for YLB, and for the little Kachi Car, the first and likely *last* EV ever built under the auspices of the Bolivian state. Although YLB even used a photo of Bolivia's new president, Luis Arce, at the wheel of the Kachi Car in publicity from 2021, the state-produced EV was not otherwise mentioned in official reports or press announcements. In fact, in the course of my own research during these years, from YLB headquarters in La Paz to the home of the Kachi Car in La Palca, the topic of EV design and production by the Bolivian state was never brought up as a serious question for either political or technological policymaking.

As we have seen, the Arce government's approach to the lithium project had already shifted markedly by this time, a shift that involved a paring back of the grand ambitions of the Morales years, a shift that, among other things, left no room for YLB's continuing experiments with EVs. In fact, the Kachi Car, which had been announced to such fanfare in 2019, had, by 2021, already been officially retired, consigned to take its place in a dusty storehouse in La Palca, perhaps next to an abandoned (and then later restored) vehicle from another era, the rusting and faded red Soviet-built transport cart nicknamed "The Red Grasshopper" (see chapter 4).

Yet if by 2021, the Bolivian government's dream of completing the lithium "productive chain" had already faded, this didn't mean that EVs weren't being produced in the country. In an extraordinary twist, they were. *Just not by the Bolivian state*. This is because during the very same years in which YLB was riding the crest of its greatest period of optimism, which was in part fueled by the global attention that came from the blockbuster announcement that the country's lithium reserves were more than double the prevailing estimates (see the introduction), a private green energy start-up was founded in Cochabamba. This remarkable company had its own vision for EV (and, later, battery) production; for how Bolivia could become a regional hub for green technologies; and for how a *private* company could contribute to the wider energy transition—a vision that had little to do with the MAS government in La Paz, YLB, or the fraught process of lithium industrialization taking place at the Salar de Uyuni.

This didn't mean that the company was opposed to the different facets of Bolivia's lithium master plan (which was itself evolving) or that it maintained a strict line of separation from YLB. On the contrary, as will be seen below, the company was more than willing to both sign agreements with YLB and associate itself with the more symbolically rich aspects of the state's resource imaginary. This complex orientation could be seen, for example, in the company's position that it hoped to *eventually* produce EVs that were powered by Bolivian batteries made with Bolivian lithium. Yet as I would come to learn, the relationship between this small company from Cochabamba and the national government in La Paz was very much a marriage of convenience. If its different linkages with YLB and Bolivia's lithium project were seen by the company to have political more than economic advantages, these linkages remained a very minor part of the company's different initiatives. Yet as will be seen, the obvious divergences between the company and Bolivia's national lithium project are themselves an important part of the wider narrative, one with implications for understanding the future course of what will likely be many, often contrasting, energy transitions.

This chapter examines the emergence of this company and, more generally, the city of Cochabamba as a center for sustainable mobility and green technologies in Bolivia. The study of this pioneering EV company and its anchorage in a Bolivian city with its own vision for the future offers lessons for thinking about how energy transitions become deeply regionalized and localized processes even in countries that have committed themselves to national policies of top-down implementation and state-controlled industrialization of critical energy resources.

In the next section, I describe the different factors that help explain the rise of Cochabamba as Bolivia's Silicon Valley of green and climate tech, before exploring in detail the history, operations, and vision of the "first EV company in Latin America." Here I focus on the company's most important product lines, which involve different kinds of EVs that are partly designed by company engineers and then assembled at the company's industrial plant in the Cochabamba Valley. The ethnography of this private company's design and production practices provides a revealing counterpoint to YLB's operations in La Palca, which I described in chapter 4. As will been seen, although these two sites of production have been conceived in starkly different political economic terms, they are shaped at the level of industrial and technological praxis by similar constraints.

Even if the first EV company in Latin America is looking to markets largely beyond Bolivia itself for its vehicles, this doesn't mean it's not also selling its EVs within Bolivia. But are Bolivians themselves interested in taking the leap to battery-powered cars, trucks, motorcycles, and delivery vans? What does the "EV revolution" look like from the perspective of even the small slice of middle class *cochabambinos* that make up the majority of the owners of the company's different vehicles? As elsewhere, the phenomenology of owning, driving, and having to justify or defend an EV is profoundly shaped by existing gender norms, political orientation, and socioeconomic class, among other factors.

The chapter concludes by considering the wider implications of the rise of a private green energy sector in Bolivia. Although companies are willing to coordinate with the national government and support a shared discourse around the critical importance of Bolivian lithium, for example, their vision for the energy transition, the role of public institutions, and the need for private capital is in striking tension with the corresponding vision of the MAS government and Bolivia's state lithium company.

AN EV COMPANY IN THE CITY OF ETERNAL SPRING

That the Cochabamba Valley and the city of Cochabamba should have become the center of a green energy florescence in Bolivia comes as no surprise, especially when viewed through the *longue durée* of productive history. Cochabamba has been nourishing wider geopolitical worlds for centuries, if not longer. At the height of the Inca Empire, which was called in Quechua *Tawantinsuyu* ("union of the four regions"), Cochabamba was a highly prized area of the empire's southern region. As the historian Brooke Larson explains, the Incas valued Cochabamba so much because it was, compared with their other mostly highland territories, extraordinarily fertile:

> In spite of [Cochabamba's] ecological diversity, it was known for its fertile, temperate valleys that caught the waters tumbling down from glacial lakes in the mountain chains to the north and west.[1]

The breadbasket of the Inca Empire became a site for massive *latifundia*, or semi-feudal estates, during the colonial period, which underpinned an enduring history marked by extreme divisions of race, class, and land distribution. The city of Cochabamba itself embodied these divisions in its residential neighborhoods, infrastructure, and labor relations. At the same time, Cochabamba became a desired location for elites of different kinds, both Bolivian and foreign. The city later gained a reputation within Bolivia and South America for having an ideal climate, and its verdant landscape became the setting for the pursuit of leisure by the city's coterie of wealthy families. Even within Bolivia, Cochabamba is viewed as a kind of Shangri-La, a valley surrounded on several sides by the towering Andes, a place of *primavera eterna* ("eternal spring"). The city is so enchanting that even the birds refuse to fly away: as a saying I heard during my first visit to the city in 1996 has it, *las golondrinas nunca migran de Cochabamba*, the swallows never migrate from Cochabamba.

The main plaza and government and cultural institutions, including the Universidad Mayor de San Simón (founded in 1832, making it the third oldest Bolivian university), are all located south of the Río Rocha, which used to form the northern boundary of the city's urban center. But over the decades, formerly rural districts north of the Río Rocha, like Queru Queru and Cala Cala, which had been part of large estates during the colonial period, were taken over by wealthy families, who built mansions along stately boulevards, a process that displaced Indigenous and peasant families farther away from the city.

Between 1915 and 1927, Simón Patiño, who was one of Bolivia's "tin barons" and one of the richest people in the world, had a European-style palace built in Queru Queru, despite the fact that he and his extended family were already living in Europe. However, Patiño suffered a heart attack, and he was never able to live in the completed Palacio Portales. In fact, no one has ever lived in the palace. Oddly, the only person ever known to have stayed the night was the French president Charles de Gaulle, who was lodged there in 1964 during a diplomatic tour of South America.[2]

The reason I dwell on the development of Cochabamba north of the Río Rocha is because it would be from these leafy and socially and geographically exclusive—and exclusionary—neighborhoods that a groundbreaking private EV and green tech company would emerge, a company deeply anchored in and dependent on the elite networks of kinship, finance, and commerce that are so characteristic of this part of the city of "eternal spring." In this sense, the beginning of a green energy transition in Bolivia itself, especially one focused on the replacement of fuel-burning cars, trucks, and motorcycles with EVs powered by lithium-ion batteries, reinforces many of the same socioeconomic hierarchies that are so notable within other national and regional energy transitions.

Even within China's massive internal EV "revolution," which is subsidized by the state at levels that far outstrip even the generous incentive programs of countries like Norway and Germany, the most affordable EVs are still only available to the country's wealthier classes. For example, the most popular EV model in China is the Qin EV300, made by the Shenzhen-based company BYD ("Build Your Dreams"). A huge government subsidy reduces its price *by almost half*; still the E300 costs about $19,000—far out of reach of most people in China.[3] As will be seen below, the founders of Latin America's first EV company were also keenly focused on the problem of cost. Even so, their own most popular model cost more than the yearly salary of, say, an urban schoolteacher in Bolivia.

The story of this company begins with a young industrial and systems engineer named José Carlos Márquez. He had grown up in Cochabamba with a father who was always working on cars, rebuilding appliances, and tinkering with machines, something that Márquez was also drawn to from an early age. His father owned a small business called Metalin, which built machines out of metal, mostly for the construction trades. The company's best-selling item was a line of portable concrete mixers. As Márquez told me, there is actually a robust market in Bolivia for these small two-wheeled mixers, which must be towed or moved by hand, because very few companies are willing to invest in the kind of large concrete mixing trucks seen elsewhere.

After graduating from one of Cochabamba's old *colegios*, La Salle, Márquez entered the department of industrial and systems engineering at one of the city's private universities, the Universidad Privada Boliviana (UPB). Although he was in his first year of studies, he continued working for Metalin, as he had in high school. As he explained, he liked the way he was able to do hands-on work with his father's company while attending classes.

However, his halcyon life was about to change dramatically. One day, without warning, his father announced that he was getting remarried and moving to the United States (his parents had been divorced for years); he and his new wife planned to settle in Fairfax County, Virginia, which has one of the largest communities of Bolivian immigrants in the world.[4] But what to do with Metalin, the small manufacturer of portable concrete mixers (and a few other items)?

His father broke the doubly bad news to José Carlos: not only was he leaving Bolivia, but he also wanted to close Metalin, which employed only four workers at the time and was, as Márquez told me, in dire financial straits. As it turned out, Metalin wasn't the only company, even in Cochabamba, that made portable mixers. Although he was only seventeen years old, Márquez's response to his father's news was wildly ambitious. He told his father not to close the company, that he, despite his age, would take over. Perhaps realizing he had nothing to lose, his father agreed, and so Márquez became the owner of a metal company while he was still a young university student.

Over the next nearly twenty years, Márquez built Metalin into one of Cochabamba's larger industrial manufacturing companies, eventually employing over fifty workers. They continued to produce portable concrete mixers but decided to expand into another line of products, a decision that would have much wider, even historic, consequences. Márquez realized that there was a major problem in what he described as the "real mining" in Bolivia—that is, "artisanal" mining by cooperatives that are associated neither with Bolivia's state mining company, COMIBOL, nor with the private transnational companies that control the largest mines in Bolivia. The problem, as Márquez explained, was that the cooperative miners had to work with the same heavy and dangerous tools as other miners in Bolivia, but they lacked the money to buy the kinds of vehicles that are used to haul mining equipment, especially in private mines. As he put it, "80 precent of mining in Bolivia is cooperative. There are no companies. No financing. That is the real mining. And they have to find a way to move a lot of things, like winches, boxes of dynamite, compressors, huge hammers."[5]

The large private mines in Bolivia use a variety of ways to transport these materials, including electric "mini-wagons," as Márquez described them, which are made by different international manufacturers—and are expensive. But the cooperatives don't have the money to buy these wagons. So Márquez decided that Metalin would design an electric transport wagon that would be both cheap and accessible to the majority of Bolivia's miners.

Márquez, who had been building and tinkering with machines since he was a boy, directed his team to figure out how the electric mining wagons worked. To do this, they spent a couple years reverse engineering all the foreign-built wagons they could get their hands on.

As he explained, they studied mining wagons made by German and Canadian companies that had highly sophisticated systems, including one that "had a small sensor that automatically stops the wagon if something passes in front of it, which makes it really safe for the miners. But this wagon costs more than twenty thousand dollars, so we knew that ours wouldn't have the same technology." In the end, they spent most of their efforts reverse engineering Chinese models and components, which is appropriate, perhaps, since Chinese companies have used reverse engineering for decades as the basis for industrial design, despite intellectual property law.

The result was the first Bolivian-made electric mining wagon, a three-wheeled vehicle that "allows one miner to do the work of many," as Márquez put it. Painted in bright orange (to enhance its safety), the wagon's metal chassis and body are made by Metalin, but the electronics, gearing, and other components are sourced from China. The battery is made by Chaowei Power, one of China's largest manufacturers of batteries for electric bikes. And it is a twelve-volt *lead-acid*, not lithium-ion, battery (see Figure 18). In between the handlebars on Metalin's electric mining wagon, Bolivia's cooperative miners can monitor the battery's charge with a gauge that uses colors (red, yellow, green) along with the corresponding Chinese characters. In the end, the cooperative miners preferred to rent, rather than buy, the Metalin wagon, but their use of the wagon, according to Márquez, did greatly improve their working conditions and occupational health.

But in 2017, the ever-restless Márquez began to think of new projects and products that would—like the electric mining wagon—both find a ready market *and* help to address a social problem at the same time. Metalin's fabrication plant was located on a large plot of land in a rural district called Colcapirhua, about twenty minutes to the west of the city center. There was plenty of room on the property to expand, even if this meant building a new plant. And then it struck Márquez: what about passenger electric vehicles? It must be remembered that this was the same year in which Bolivia's state lithium company had been founded. As we have seen above, although YLB and the MAS government occasionally gestured to the possibility that the state might someday also make EVs, as Márquez understood, this was never taken seriously.

FIGURE 18. Metalin's electric mining wagon, Cochabamba, 2022. Photo by author.

At the same time, the global buzz around the EV "revolution" was becoming more pervasive, something that Márquez also noticed. He started to consider the idea in detail. As he put it, "I thought that I had a very good understanding of the principles of electromobility." A car would be just a more complicated version of the electric mining wagon. *Once we had figured out how to integrate the different components, the battery, the controller, the motor, and the accelerator, we started to look [more seriously] at the issue.*

But how does a Bolivian metals manufacturer in Cochabamba launch an EV company? Where does one start? Márquez knew right away that he needed help, someone from outside Metalin who could manage all the necessary steps, from the negotiation of international supply contracts to fundraising for the new company to the enthusiastic promotion of a homegrown EV company in Bolivia—an initiative that Márquez knew was going to be treated as having very little chance of success, one, moreover, that didn't have either the political or financial backing of the Bolivian state.

Márquez turned to an old friend for help, Carlos Soruco Deiters, a thirty-seven-year-old commercial lawyer who managed his own law firm, Soruco & Associates. The two of them had known each other for almost twenty years and ran in the same elite *cochabambino* social and economic circles. Moreover, Soruco had done considerable legal work for Márquez and Metalin over the years, work that involved commercial agreements, intellectual property applications, and supply contracts. More than anything, however, Márquez thought that his old friend Carlos Soruco would be an ideal partner in the risky enterprise for a more intangible reason: much more than Márquez himself, Soruco hailed from one of Cochabamba's storied families, one that had exercised significant power within the city's (and country's) political, legal, and cultural life for over a century. Given the profound uncertainty at the heart of the venture, Márquez thought it would help to have such a well-connected partner, someone with social, business, and family relations across the kinds of networks that the new EV company would need to access to have any hope of success.

On his mother's side, Carlos Soruco's family (Deiters) comes from Germany. His maternal grandfather immigrated to Bolivia from Bremen, where his great-grandfather had been the mayor of the German city. Soruco's father, Jorge Soruco Quiroga, is also a lawyer and key figure in Cochabamba's tightly knit business community, who also served at various times in national political roles, including a term as the national secretary of justice during the first government of Gonzalo Sánchez de Lozada (1993–1997). Soruco's uncle, Juan Cristóbal Soruco Quiroga, is a distinguished journalist and, until 2018, the longtime director of *Los Tiempos*, the main newspaper in Cochabamba, which has been published since 1943.

The list of notables in Soruco's family extends back into the early twentieth century, when his great-grandfather, Ricardo Soruco Ipiña, also a lawyer, was a member of the Bolivian congress from Cochabamba. One of the intellectual leaders of Bolivia's small Socialist Party and a strong advocate for labor unions, Soruco Ipiña was the only member of the Bolivian Congress from any party to publicly denounce one of the most notorious atrocities in Bolivian history, the Jesús de Machaca massacre of 1921.[6] In 1954, two years after the National Revolution, Soruco Ipiña (then seventy-five years old) was awarded the Order of the Condor of the Andes, the highest honor given by the Bolivian state.

Carlos Soruco himself grew up in Cochabamba, where he attended a number of private schools, including the German School "Federico Froebel," which is part of a network of German schools inspired by the pedagogical theories of the nineteenth-century German educational reformer Friedrich Fröbel (who invented the kindergarten). Soruco then studied law at the Universidad Católica Boliviana, San Pablo, also in Cochabamba, after which he was licensed to practice law. However, instead of settling into a legal career right away, Soruco decided to move abroad for graduate school, a decision that would have major implications not just for Soruco's career but for the course of Bolivia's own incipient green energy transition.

Leaving his family and friends in Cochabamba behind, Soruco moved to the Netherlands, where he enrolled in a master's program in international economic and business law at the University of Groningen. Why did he decide to study in Groningen? As he explained, the program was apparently the "top" one in Europe, even though he "didn't even know where Groningen was." Apart from attending classes and participating in the wider life of one of Europe's iconic university towns, Soruco wrote a master's thesis on the MAS government's nationalization of the country's oil and gas industries in 2006. He argued that the expropriations actually complied with international law, since the Morales administration offered "just compensation" to the private companies and the nationalization could be justified as a public good.

But apart from his classes and research, it was something else that he experienced in Groningen that would have the most lasting impact on Soruco's worldview and later trajectory: the famous predominance of bicycles in the Netherlands, a tradition that is even more conspicuous in university towns like Groningen and Leiden. One of the first things he did after arriving was to buy a bicycle, which he rode everywhere. *In Groningen, you do everything on a bicycle, you seem to spend your entire life on it. You are always riding from one place to another, to the university, to the disco, everywhere.*

When he finished his master's degree in 2008 and returned to Cochabamba, he experienced something like a reverse culture shock, at least with regard to transportation. *I got back and couldn't believe it. It never rains in Bolivia, or, at least, the rainy season is relatively short. The distances around town are also fairly short, and it's flat, which means it's an ideal city for bicycles. But where were the bicyclists? And then I remembered from when I was a child, my parents used to cycle, you might even say that Cochabamba used to be a cycling city. But not anymore.*

Yet Soruco was not just frustrated that an ideal cycling city lacked either the urban infrastructure to support cycling as a form of public transportation or sufficient numbers of cyclists willing to brave the streets of the city regardless—he was angry. As he put it, he would arrive at his law offices on Avenida América in the heart of Queru Queru after commuting along streets chocked with cars, *micros* (city buses), *minibuses* (passenger van-sized buses), *trufis* (cars operated like buses), and taxis, "red with fury."

So he decided to do whatever he could to make Cochabamba more like Groningen. These were also the years in which the global cycling activist movement Critical Mass was becoming more visible. Using both his legal training and deep connections with Cochabamba's business community, city government, and *Los Tiempos* (which his uncle directed at the time), Soruco launched a public campaign for Bolivia's first "bicycle law," which would create a legal foundation for transforming the city into a zone where people used bicycles as a primary mode of transportation. But despite his capacity to mobilize support among key sectors of the city's elite, Soruco's vision ran into obstacles at every point.

There were questions from the city government about the costs of building an urban cycling infrastructure. There were concerns from local businesses about whether cyclists would be able to transport goods and otherwise travel around the city. Although Cochabamba is relatively compact, it isn't *that* compact. And although it sits in a relatively flat valley, compared to, say, La Paz, it is not completely flat like Dutch cities. Many of its neighborhoods are in the foothills, further complicating the vision of Cochabamba as a cycling city. And then there was the more basic problem, as Soruco soon discovered, that most people in Cochabamba were either completely indifferent to the idea of cycling as a regular mode of transportation, or even opposed to it. People who could afford to own cars preferred to drive them. And the rest were quite content to use the hodge-podge of highly polluting, yet public, transportation that they had always used. Many people were also accustomed to walking regularly, even across long distances.

But Soruco kept up the struggle for almost a decade. He became known in town as a cycling activist, even as he maintained his commercial legal practice.

It was a terrible struggle. The municipal government agreed to study the issue, but that dragged on for years. But then something changed. I managed to convince the Cochabamba Chamber of Commerce to hold a press conference with researchers from the Universidad Católica Boliviana [his alma mater] *who sup-ported the idea* [of a bicycle law]. *A professor in the department of communications designed a course in political marketing around the struggle* [for the law]. *Eventually, the city government started to pay attention.*

Soruco's efforts were finally rewarded in 2017, when the city government passed Law 221, the "City Bicycle Law." Using ecological, public health, and economic justifications, the law establishes a civic right to bicycle in Cochabamba while also promising to undertake a series of reforms to the city's infrastructure in order to increase cycling safety and visibility, including the promotion of cycling as a sport. But as Soruco acknowledged, even though the passage of the law was a milestone for the city, its actual impact over the subsequent years was minimal. The city did manage to construct some concrete bike lanes, although most of them pass through the center of parks or mediums rather than along the sides of city roads. And by 2022, five years after the law took effect, there had been no appreciable change to the city's transportation patterns. Very few people were willing to cycle along roads that were as clogged as ever with a chaotic flow of cars, *micros, mini-buses, trufis*, taxis, motorcycles, and diesel-belching trucks.

Yet even if the decade-long campaign for the city bicycle law was, from a practical standpoint, a failure, it had another unexpected outcome: it made Soruco a somewhat unlikely advocate for urban sustainability and energy transition. This new profile partly explains what happened next. Soruco was already, as we have seen, doing commercial legal work for Metalin, including the negotiation of supply contracts for the company's electric mining wagons. José Carlos Márquez, Metalin's owner, was looking for a new major project. He had been following the public activism of his longtime friend and lawyer with interest over the years. He then told Soruco about his idea: he wanted to start a new company, one that built and distributed a variety of EVs: cars, motorcycles, delivery vans, and scooters.

What did Soruco think? As Márquez explained, "I remember coming to Carlos the first time and telling him my plans. 'I want to build electric cars.' Carlos just looked at me for a moment and then said, 'Cars?' He was incredulous at first." But Soruco agreed to strategize with Márquez and to at least consider the different implications. What kind of cars? What would the legal and financial requirements be to start a car company in Bolivia? Where would they build these vehicles? As he had done with the electric mining wagons, Márquez wanted to build EVs that would have a market within Bolivia, and perhaps in other countries with similar needs and limitations.

> *As I told Carlos, I had already decided on the kind of EV I wanted to design and build. It was already going to have a certain size. Why? Because it had to be cheap and in order to be cheap it had to be small. Because for bigger, heavier cars, you need more expensive batteries, a bigger engine. I also knew that our EVs had to cost less than $7,000. Why? Because at the time there was a popular [fuel-burning] car for people here made by Suzuki that cost $9,000 brand new, even cheaper in a used model. So for reasons of cost, I knew that we would have to make small EVs that were more affordable than the cheapest Suzukis.*

Despite his initial skepticism, Soruco eventually decided that the idea might work. He agreed to be part of the project, on the condition that he could continue managing his commercial law practice. That settled, Márquez and Soruco founded their EV company in 2018, which they called Quantum Motors. But once they had cleared the initial regulatory hurdles, they confronted a number of problems, including securing financing for a start-up company that planned to make a very unlikely (for Bolivia) product. Given his long-term connections with Chinese suppliers, Márquez suggested that the two of them should travel to China to try and put together a feasible production model. They spent more than a month visiting business associates, factories, and cities such as Beijing, Shanghai, Guangzhou, Wuxi, and others. Soruco, in particular, was ecstatic about what he saw and experienced in China.

> *It was incredible, amazing! The cities are huge, much bigger than you can imagine. And they are more developed than cities in the United States, even more than in Europe. It would blow your mind! I've lived in Europe, in the United States, but China is something beyond belief. It was in everything, the road infrastructure, the cleanliness, the safety, the technology.*

More to the point, they managed to secure a number of licenses to incorporate components from Chinese EV companies. The cofounders of Quantum decided that they couldn't afford to design and produce completely new EVs in Bolivia, but they also didn't want to simply import fully assembled Chinese EV models. Because of Márquez and Metalin's expertise and production capacity with metal, they decided they would fabricate most of the metal parts for Quantum vehicles in Cochabamba. But they would have to import most of the other vehicle parts from Chinese suppliers, including electronics, motors, and, most importantly, batteries and related peripherals (like the BMS, or battery management system). They settled on a design for their first EV that was inspired by the Chinese models that were common on the roads of the cities they had visited.

Here, they faced another major decision: once they had decided on a design, would they build the body of the car itself in Cochabamba? As Soruco explained, to make the composite plastic doors, trunks, roofs, hoods, etc., for the EVs, one needed specifically designed molds. With molds in place, a new car company can then begin producing the bodies of vehicles at something like an industrial level, assuming the company has the right materials, machinery, and production space. But as Soruco explained, they were told that they would have to purchase the molds from their Chinese partners for between $3 and $5 million. Because their initial funding didn't come anywhere close to these amounts, they realized that they would have to simply import the vehicle body parts along with everything else.

And what about the all-important batteries? Would the first Quantum EVs use *lithium-ion* batteries? As Márquez described it, they decided to use lead-acid battery systems for their first models, despite the fact that they would perform considerably less well in terms of range, durability, charging speed, and so on than lithium-ion alternatives. (Lead-acid batteries are also not usually associated with the energy transition.) Why? Because they were working under the constraints of a target price of $7,000.

> *To get to $7,000, you need to use fewer materials, cheaper materials, smaller motors, smaller batteries, everything. When we entered the market* [in 2019], *lithium-ion batteries were much more expensive than they are now* [2022]. *Much of the cost of an EV is in the battery. When we launched, we did the calculations. Based on the price per kilowatt-hour, our EV* [with a lithium-ion battery] *was going to cost about $10,000. No one was going to want to buy it. So we began initially with lead-acid batteries in our models.*

With the law offices of Soruco & Associates doubling as the new corporate headquarters of Quantum Motors, the company built a large covered production space adjacent to the existing site of Metalin in a stretch of *campo* to the west of Cochabamba, where the first EVs made in Bolivia would be produced among fields of corn and the occasional dreaded *winchuka*, an "assassin bug" species, which is a carrier of the incurable Chagas disease, common to the Cochabamba Valley. Once they had secured the many different parts and components—which they import from their Chinese suppliers via

FIGURE 19. Quantum factory, Cochabamba, 2022. Photo by author.

the northern Chilean ports of either Iquique or Arica—they set up a factory of sorts with spaces for painting, assembly, repair, testing, and metal fabrication, and so on (see Figures 19, 20, 21, and 22).

At the same time, they rented the ground floor of an office building not far from the corporate offices, where they installed a showroom for their vehicles. With Márquez responsible for all things related to engineering and production and Soruco responsible for Quantum's legal, commercial, and marketing strategies, they hired employees for both the factory and showroom. Although Soruco wouldn't go into details about the company's initial financing, he explained that Quantum received funding from a small group of mostly local investors. Again, Soruco's deep roots in Cochabamba's interconnected business, legal, and social networks proved critical for securing the company's start-up

FIGURE 20. Quantum workers apply paint to an EV, 2022. Photo by author.

capital. By 2019, everything was in place to begin making their first cars.[7] But would the company be able to convince people—at first, only in Cochabamba—to make the transition to EVs, to actually buy an electric car made in Bolivia?

ELECTRIC, EFFICIENT, ECOLOGICAL, ECONOMICAL

Yet as we have seen with other institutions that are part of the wider story of lithium and the energy transition in Bolivia, Quantum was likewise buffeted by the political, social, and public health crises of the period. The company opened its showroom with considerable pomp and media attention in September 2019. Soruco describes what happened next:

FIGURE 21. Workers assemble an EV at the Quantum factory, 2022. Photo by author.

Right after our big opening, there were the elections [in late October] and all hell broke loose almost immediately. Here in Cochabamba, everything had to close during the protests, we had to close the factory, the showroom, everything. The whole city was shut down by strikes, blockades—it was complete shit.

When were you able to actually begin operations?

Well, after all of these political and social problems, then the pandemic hit. Imagine, we had signed contracts with suppliers from China for most of our parts and then everything came to a halt just as we had begun receiving shipments. Our partners [in China] told us, "All the factories here are closed, no one is working." And then we had shipments that were already coming from China when all the borders were closed. The ports in Chile were shut, so our containers were just stored for later. I mean, it was really impossible to have found a worse moment to be born [as a company], but we survived somehow.

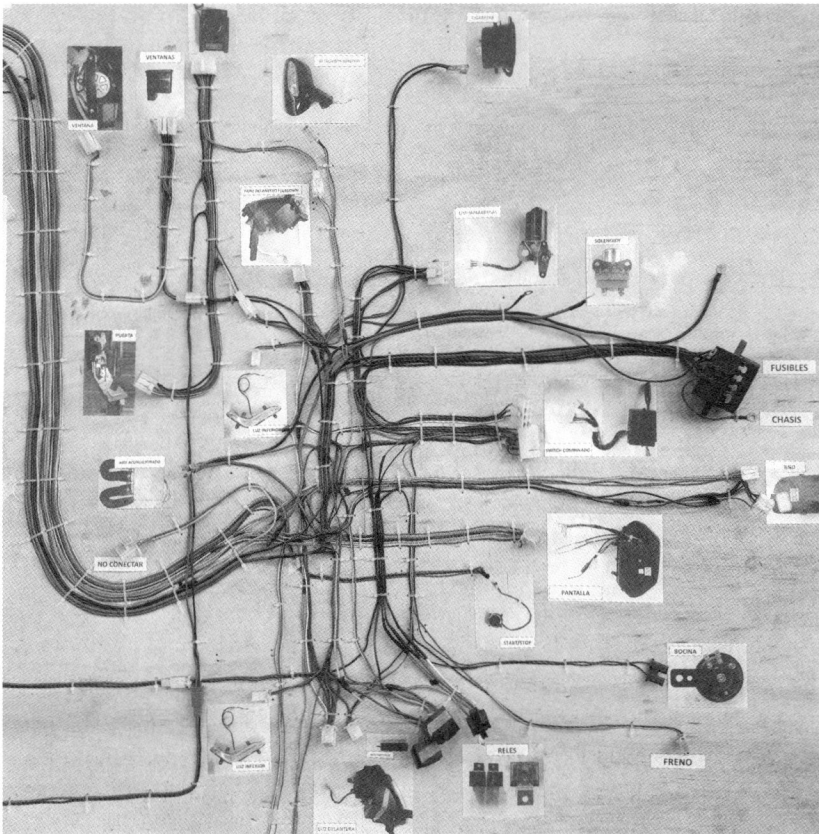

FIGURE 22. Handmade wiring diagram, Quantum factory, 2022. Photo by author.

Although Bolivia remained under various restrictions, even after passing through the worst of the pandemic, Quantum was able to restart production and reopen its showroom by early 2021. The company was more than ready, including with a multifaceted branding and marketing strategy that sought to position Quantum as a new player in the wider green energy transition. They adopted a number of marketing slogans and logos. One slogan, which was used on brochures and posters, proclaimed: "Quantum Electric Vehicles—It's the moment to change the world! Maintenance-free, gasoline-free, using less energy than a clothes iron."

Another slogan, which they emblazoned on the sides of some of their early EVs, was "electric, efficient, ecological, economical," which Soruco described as the "four Es," the principal values of the company. I thought I understood the meaning of three of the four Es, but what about "efficient"? I was picked up in one of the cars by Quantum's head of communications and driven to the showroom, my first experience in the little EV. As Soruco explained, "Well, you could have driven here just as easily in a gigantic Range Rover or something like that. But why? You don't need anything bigger [than a Quantum] to get around the city. It's super-efficient, it takes you wherever you want to go in town, no pollution, no money spent, so it's efficient in many ways."

The company had to come up with a name for its EV models and they chose "E" yet again, which stands for Explorer. The first Quantum Explorer was treated as a prototype, so the second model, and the first they actually marketed and sold, was the E2. By 2022, they had moved on to the E3 and E4, although they continued to offer maintenance on older E2s at the factory in Colcapirhua. In terms of size, the E3 and E4 are identical. Quantum markets the EVs as "three-person vehicles," with a single seat up front for the driver and a second rear seat that is theoretically for two people (including two sets of seat belts).

But as Soruco explained to a journalist from the *Wall Street Journal*, something he also told me after I mentioned that there didn't seem to be enough room for two adults in the back seat, the three passengers "can't be very fat." As the *WSJ* article—which is entitled "Look Out, Tesla, There's a Really Tiny Competitor in Your Rearview Mirror"—puts it, "High in the Andes, the electric-vehicle revolution has arrived. It's moving no faster than 35 mph and can be a bit claustrophobic."[8]

The decision to make their EVs "tiny" was quite intentional and part of Quantum's broader vision, one that is partly shaped by economic concerns, as we have seen, but also shaped by what Soruco and Márquez saw during their transformative business trip to China. In this vision of the energy transition, individual transportation, especially in cities, takes place through *electro-micro-movilidad*, which involves the use of very small cars and other vehicles (like three-wheeled cargo vans and electric motorcycles) that are just big enough to get people around relatively short distances, widely affordable, and carbon-free. As Soruco explained,

> *Electro-micro-mobility in China is everywhere, it's huge. I think combustion motorcycles, for example, are actually banned in all the major cities. But the use of micro-EVs is going to break the hold of bigger cars, so we have faith in this type of mobility. And micro-EVs are also the future of car sharing. There are already so many tiny shared cars in Spain. We want to introduce this idea to Latin America.*

The differences between the E3 and E4 have mostly to do with the size of the engine: the E3 comes with a 3,000-watt engine, while the E4's engine is rated at 4,000 watts.[9] However, despite the 25 percent increase in power, which makes a difference for customers who live in the foothills north of the city center, the other performance characteristics are more or less the same. Both models have a maximum speed of 55 km/h and a maximum range on a single charge of between 50 and 55 km—that is, with a lead-acid battery. And even though the cars can be purchased directly with a lithium-ion battery (assembled by Quantum, also from Chinese components), many customers choose a different strategy: they buy their new EV with a cheaper lead-acid battery with the hope that they will be able to afford to retrofit the car with a more expensive lithium-ion battery after a few years. However, even though a lithium-ion-powered Quantum has a range that is almost double (100 km), several owners told me that the trade-off wasn't worth it because 50–55 km on a single charge is more than enough to cover their normal driving during the day.

By late 2022, Quantum had managed to sell almost four hundred of their EVs. The manufacturing process at the factory in Colcapirhua is made-to-order. They don't begin producing a new EV until a customer has made a series of choices in the showroom, including about engine size, battery type, and color. Under this system, they were producing about twenty EVs per month under conditions that were a bit difficult to describe analytically. After conducting research over several years at the factory, during which only small changes were made to the process, I began describing it as "artisanal," meaning a form of production that was not automated but tailored to the tastes of each customer (within limits). And the quantity of EVs produced also seemed to suggest an artisanal approach.

However, when I mentioned this emerging analysis to Soruco, in particular, he became quite sensitive, even a little angry, despite the fact that what I wanted to emphasize with "artisanal" were things like attention to detail and the relationships the company was forming with the drivers of its EVs. Perhaps given the fact that Quantum has big plans for expansion, which will require significantly more capital investment and industrial capacity, the term "artisanal" understandably didn't capture their vision for the future. Moreover, artisanal production in Bolivia is associated with rural economies or traditional handicrafts. In the end, we agreed that I would describe the scale of Quantum's production as "hybrid": part artisanal and part industrial.

But given that Quantum had become an icon in Bolivia for a specific approach to the energy transition, one that was gaining attention from global media and energy analysts who were eager to draw contrasts between the *private* start-up in Cochabamba and Bolivia's *state-controlled* lithium company, what was the actual relationship between Quantum and YLB? Márquez and Soruco were keenly aware of the fact that a structural tension existed between their EV start-up and the country's lithium master plan, even though—as we have seen above—the MAS government's gestures over the years to state-built EVs were never really taken seriously at the level of state financing or policymaking, something that became even clearer after Luis Arce's administration resumed stewardship of the lithium project in late 2020 (see chapter 3).

Nevertheless, Quantum knew that it had to walk a fine line, that it was, despite its ambitions, vulnerable to different forms of potential political and legal scrutiny from La Paz. Its publicity campaign went to great lengths to emphasize the ways in which the company was deeply committed to the country's wider lithium strategy. For example, although it sourced its battery components from Chinese suppliers, Quantum signed an agreement with YLB to procure their battery cells, even though YLB had no plans to ramp up its operations in La Palca to produce battery cells beyond a "pilot" level (see chapter 4).

Neither Márquez nor Soruco were ever critical or dismissive of YLB and the wider political economic model it represented, even in off-the-record discussions. But the actions of Quantum—its reliance on private capital; the rollout of its business strategy, which eventually included the opening of showrooms in El Salvador, Peru, and Paraguay; and the beginning of negotiations to build a second Quantum factory in Mexico—spoke volumes.

From the side of YLB policymakers, Quantum was viewed largely with indifference, mostly because the state company was much too busy reckoning with its mounting debt and laboring under a kind of permanent institutional uncertainty. Moreover, the shift in approach toward what I have described as "flexible extractivism" (see chapter 3) under the pragmatic technocrat Arce meant that the government's attention became narrowly focused on the industrial production of lithium carbonate. As we have seen, this had the effect, among other things, of marginalizing the state's battery operations and pushing Bolivia even further from its original ambitions to vertically integrate the lithium "productive chain." Under these wider conditions, it was not surprising that YLB took what could be construed as a paradoxically

FIGURE 23. Proud Quantum owner poses with her EV, Cochabamba, 2023. Photo by author.

laissez-faire approach toward Quantum, especially given that the small company in Cochabamba actually represented a radical challenge to much of the ideological framing of Bolivia's slowly unfolding lithium project.

SÍ, SOY ELÉCTRICO

But if Cochabamba had become something like the locus for an alternative energy transition in Bolivia, with Quantum as its avatar, what did this transition look like to *cochabambinos* themselves? Leaving aside the stalled initiative to transform Cochabamba into a cycling city à la Groningen and the work of green energy NGOs like Energética, how did people in the city respond to the call to be part of the vanguard of Bolivia's own nascent EV revolution?

My first attempts to answer this question ethnographically led to a profound sense of skepticism. During an early period of research in Cochabamba, I decided to document the ebb and flow of transportation in the city, to try and understand something of the phenomenology of mobility. Despite the city's topography, which, at least within the more central districts, is relatively flat and crisscrossed by numerous walking paths and sidewalks, the vast majority of mobility is motorized. In contrast and despite the fact that the topography of La Paz couldn't be more different, with its steep streets and considerably higher altitude, central La Paz is very much a pedestrian zone, even though the streets themselves can still become clogged with traffic.

To capture this preexisting culture of mobility in Cochabamba, I shot a short ethnographic video of the busy intersection where Quantum decided to open its showroom. Standing at the edge of Hipermaxi supermarket's large parking lot, which was filled with the (fuel-burning) cars of shoppers, I filmed the ceaseless cacophony of horns and the never-ending rush of passing vehicles. The Quantum showroom sits in the background, one of its models parked outside, ready to be taken for a test drive. As the river of combustion engine vehicles of all types flows in every direction, Quantum's main slogan is nevertheless still visible from the large window: "Electric, efficient, ecological, economical."

But despite these stark contrasts, I came to learn that there *is* a growing interest in electromobility, micro- or otherwise, in Cochabamba, an interest that is closely tied—as elsewhere—to socioeconomic class. Even though Quantum did manage to meet its stated targets for affordability—the E4 had a sales price of $6,200 and the E3 listed for $5,900—the vehicles were still far beyond the reach of many people in town, not to mention the hundreds of thousands of people living in Cochabamba's peri-urban *barrios* and villages throughout the department. Yet the wider messages of sustainable mobility and the importance of transitioning to green energy are embraced by many middle-class *cochabambinos*, including the city's large student population, who rent e-scooters to get around the city center from one of several private companies.

The perspectives of those who actually *have* made the transition themselves to EVs reflect, as elsewhere, a diversity of reasons for having made the change and experiences with the cars. For example, when I met her in 2023, Maria was a fifty-nine-year-old psychologist who worked in a private high school.[10] In the late 1960s, during the regime of General René Barrientos, Maria lived in the US for several years with her family in the northern Utah town of Logan, where her father did graduate work at Utah State University in agricultural engineering. A few years after they returned to Bolivia, her own driving history indirectly began when the family bought a 1975 Nissan Patrol 4×4, which she referred to as the "Jeep." She later learned to drive in the Jeep, and then her parents gave it to her when she was in her twenties.

By 2020, Maria still owned the now-forty-five-year-old Jeep; in fact, it was the only car she had *ever* owned. As she put it, the stories she could tell about her experiences with auto mechanics over the years could themselves be part of an academic study. Eventually, however, she grew tired of the constant upkeep and cost of maintaining the ancient Jeep. She was also becoming more attuned to the problem of climate change and the importance of moving away from fossil fuels. But she was worried about whether she could afford a new car after so many years. At first, she considered buying an electric motorcycle, but her mother, with whom she had a close relationship, tried to talk her out of it, mostly for safety reasons. At the same time, she had read in the newspaper about Quantum and was captivated by the idea of what she called their "little cars."

Despite her curiosity, she made her decision: she would buy an electric motorcycle. But before she could do it, her mother intervened with a surprising offer: if Maria really wanted a Quantum EV, she would give her $5,000 in order to make the purchase. As Maria explained,

> *My mother heard from my brother that I had decided on the motorcycle. She called me immediately. I told her, "If I buy the car, I'm a little afraid, because I'm on the edge financially and don't have any backup." "Ok," she tells me, "I will give you the money for the little car and you can do what you want with it." Wow! So, really, it was my destiny to have a Quantum.*

With the gift money in hand, Maria became one of the first owners of a Quantum EV. She bought the E3 with a lead-acid battery. By 2023, she had replaced it with a lithium-ion battery. As she told me, it took her quite a while to know what to do with her new "micro" EV. Although she had been driving for decades, it was difficult to imagine such a small car as the replacement for the Jeep. At first, she drove it sparingly, taking a series of nervous "test drives" over about a month. As she put it, she and her new EV had to learn to "trust each other." She eventually got used to all the differences in size, speed, and function. When this happened, the Quantum became her trusted companion, although one that drew different kinds of attention.

She told me that she rarely manages to drive from one place to another in Cochabamba without something happening, usually taking the form of a kind of intense—and even dangerous—inquisitiveness. People drive too close to her to get a better look at the tiny EV; they make comments through open windows at red lights; and she catches people touching or even rocking her Quantum in parking lots, as if they want to know whether it will tip over. But at least in the middle-class neighborhoods of the city where she spends most of her time, the most common question she receives is whether she would recommend the EV to others. How does she respond? "Absolutely."

Victor's experiences as an owner of a Quantum EV were equally idiosyncratic. In his late-twenties when I met him, he had studied chemical engineering at Cochabamba's major public university, the Universidad Mayor de San Simón. While he was a student, both of his older brothers moved to France for work and they told him about EVs, especially the Teslas they had seen on the streets of Paris. Coincidentally, when he graduated in 2018, his first job was at La Palca, where he worked as part of YLB's battery development team. While he was living in Potosí, his mother called him one day: "'Victor, you won't believe it, there are already electric cars in Cochabamba, I just saw one!' Are you sure it was electric? I asked her. 'I think so,' she said. 'And it was very small, a little car.' I started to laugh and then she asked, since I was working for the state lithium company, 'What are they [EVs] doing here?'"

When Victor left YLB in 2021, he moved back to Cochabamba to work in private industry. Although he already owned an aging Nissan SUV, he decided that he simply *had* to have one of the first EVs made in Bolivia, given both his area of scientific expertise and his commitment to the wider cause of sustainability. Like Maria, he decided on the E3. And also for reasons of cost, he opted for a lead-acid battery. His family and friends teased him for the irony of his choice: "'Hey, Victor,' they said to me, 'Why would you buy an EV with a lead-acid battery? We thought you were working on lithium? What's wrong with you?'"

But even a lithium-ion battery scientist can find a way to justify *not* buying an EV with a lithium-ion battery, that is, beyond the obvious financial reasons. As he explained:

> I know it might seem strange to say, but lead-acid batteries are good batteries too. They have a long life if you use them properly. For example, you need to

use the [EV] at least twice a week if you don't want to degrade the battery. And you also need to make sure to disconnect the cable once it's done charging, which also helps to preserve the battery. For now, I can't really afford to upgrade to the lithium battery. I would have to sell the Nissan, so that's something I'm looking into for the future.

Speaking of the Nissan, I asked him about the differences between his two cars, the larger gasoline-guzzling SUV and the Quantum, with its top speed of 50 km/h and its maximum range on a single charge of 50 km. He expressed some remorse for holding on to the Nissan, despite his belief in the importance of the wider energy transition. But then he drew an astonishing contrast, one that I had never heard before: according to him, the small Quantum was actually *safer* than the larger Nissan. Most Quantum owners had told me that they had felt menaced at one time or another on the roads of Cochabamba as they tried to make their way among the much faster and much bigger vehicles. How could the Quantum be safer?

As Victor explained, it had to do with a phenomenon that's not unique to Cochabamba: the high incidents of theft, in which parts are stolen from parked cars. As in other countries, catalytic converters are a particular target. Thieves can earn hundreds of dollars from the palladium and platinum used in catalytic converters, which reduce pollution emitted by fuel-burning cars. But as Victor put it,

With Quantums, this never happens. You can park it anywhere and it is completely safe, no one will try and steal parts from it.

Why not?

I don't know, I think it's because there is no market for these parts. Who would buy them? With the other cars, they take everything here, electronics, computer systems, anything. Maybe in the future, if more people buy them [Quantum EVs]*, this will change. I hope not* [laughing]*!*

Finally, Roxana, another Quantum owner, admitted that her decision to buy an EV wasn't based on a deep concern for the environment or the need to transition to cleaner energy technologies. As the twenty-eight-year-old accountant put it, "I was never really a fan of all that environmentalist stuff." I asked her if this was for political reasons. *No, not really. I don't know, it's just not something that ever caught my attention. Ok, I keep garbage in my backpack, I don't throw it in the street, things like that, but nothing extreme.*

In 2022, because her work was going well and she had earned a salary for a few years, she decided that it was time to buy a car. As she explained, "I had seen a number of articles in the newspaper, and advertisements, about Quantum. You would see articles about how a 'Bolivian company manufacturers electric cars,' and things like that. But I just wanted a car, not necessarily an electric car." As with other Quantum owners, the circumstances that led Roxana to hers were also marked by a certain randomness (what Maria otherwise described as "destiny").

Roxana tried at first to buy a used car, but the owner of one she really wanted "seemed to disappear" every time she wanted to have it examined by a mechanic. She explains what happened next:

> I told my parents about the problem with the used car, because I still live with them. Then my dad said, "Why don't you look at a Quantum? Because at the end of the day, why do you need a car? You will use it to go back and forth to work. If you want to go somewhere else, that's no big deal. Let's say you want to go on a big trip to La Angostura with your friends, you can just take our car." I agreed, and so I bought a car from Quantum.[11]

Roxana bought an E3—again, also with a lead-acid, *not* a lithium-ion, battery—and named her new EV "Katy." Why? "Because of the license plate: _ _ _ KTY."

> My little car is like a new member of the family now. I tell my friends, "I've come in the Katy." Or, my friends will ask me, "Where's Katy?" [laughs]. My Katy has become a good friend, and I've tried to personalize her. I'm a Harry Potter fan, so I put a sticker on the back that says, "Hogwarts alumni."

However, not all of Roxana's experiences in Katy have been pleasant. As the energy scholar Cara Daggett has argued, the transition to EVs is infused by existing gender norms, especially those in which cultures of driving enable expressions of masculine aggression, something Daggett describes as "renewable masculinities."[12] This is similar to what Roxana and other female Quantum owners have reported. As Roxana explained, as a young woman driving an objectively small, relatively slow electric car, she sometimes feels vulnerable, especially on the longer boulevards of Cochabamba, where cars drive faster. What does she do to protect herself? *I've learned to be very careful; I usually stay as far to the right of the road as I can, but even then, it seems like other drivers sometimes don't respect me.* Have these experiences ever made her reconsider her choice to buy a "micro" EV? *No, not really. One manages. In fact, I've started to think about making the change to a lithium-ion battery, I've started saving money . . .*

CONCLUSION: "THE BEGINNING OF A DIFFERENT BOLIVIA"

The remarkable rise of Quantum, an unlikely EV start-up from Cochabamba that imagines its cars changing the world by using "less energy than a clothes iron," is an essential moment in the wider history of Bolivia's lithium ambitions, even as it unfolds along an alternative track, one that must be understood as a critical counterpoint to the government's still-evolving lithium strategy. By way of conclusion, I want to reflect on how the status of Quantum contrasts with the different dimensions of the state's much more visible lithium project. At the same time, these obvious contrasts do not mean that there are no points of convergence between the vision of Quantum and that of YLB. At the level of discourse, if not political economy, Quantum embraces some of the same ideological aspirations as Bolivia's state lithium company, even if they are expressed quite differently.

It is important to recognize the fact that the emergence of a private EV company in Cochabamba took place at a high point for Bolivia's state lithium project—the culmination of a decade of unprecedented levels of state investment, international negotiation, technological development, and global interest in Bolivia's lithium reserves bordering on obsession. Yet even before the course of the lithium project in Bolivia took a dramatic and transformative turn in the weeks and months around the 2019 national election and ouster of Evo Morales, the foundations of the original lithium master plan were already coming apart.

With the dream of creating a lithium-ion-battery industry in Bolivia more or less moribund by this time, at least behind the scenes (see chapter 4), even more remote links in the idealized lithium "productive chain," specifically the manufacture of EVs by YLB, were detached even faster. Although YLB never had an official policy in relation to Quantum, its presence wasn't viewed as particularly threatening; if anything, officials in both YLB and the wider Ministry of Hydrocarbons and Energy were willing to let things play out with Quantum, since closer ties—perhaps even under legal compulsion—could always be forged later.

But this is not to say that actors on both sides—YLB and Quantum—weren't keenly aware of how much the public image and political economic vision of the private company in Cochabamba clashed with Bolivia's state-controlled lithium project. If both the MAS government and YLB fore-grounded the ways in which the country's lithium project would contribute to a politics of "productive sovereignty" while generating enough revenues to end the country's reliance on gas, Quantum viewed its national role, and contribution to the wider energy transition, quite differently.

There is no question that the cofounders of Quantum were part of a highly influential community in Cochabamba that has a long history of privileging the place of private industry and investment over state intervention in the economy. In comparison to the wider population even of the city of Cochabamba, not to mention Cochabamba Department, this community has always been quite small, exclusionary, and dominated by families inter-connected by marriage, social ties, and business arrangements. As the massive mobilizations against the privatization of water in Cochabamba in 1999–2000 proved, a period known as the Water War,[13] the majority of *cochabambinos* have a more nuanced and contextual perspective on the place of private industry in their lives.

But for José Carlos Márquez and Carlos Soruco, if not necessarily for Quantum's rank-and-file employees, there was no doubt that Bolivia's wider lithium ambitions, including the production of batteries and EVs, would be much better served in the hands of private Bolivian companies, private tech-nology researchers, and partnerships forged through private trade relation-ships. In this sense, Quantum is aligned with the political economic thrust of so much of the global green energy transition, which is manifestly *not* taking place through state-controlled industry, despite the heavy hand of the state in setting policy, establishing regulatory frameworks, and enforcing (or not) environmental laws. And even though the Chinese companies that con-trol much of the global lithium, battery, and, increasingly, EV markets, are tightly monitored by the Chinese state, they operate in a kind of political economic gray zone in which they function very much like other major capi-talist entities.

However, although Quantum is not, like YLB, working "for the benefit of all Bolivians"—revenue generated from the sale of its EVs will obviously not be redistributed as social benefits—this does not mean that the company is opposed to energy or technological nationalism of the kind that is so often invoked by the policymakers at the center of Bolivia's lithium project. But Quantum's vision for how its products will advance the national interest expresses a nationalism of a very different kind.

As Márquez puts it, in promotional material announcing the company's newest venture, the launch of the first industrial lithium-ion battery factory in Latin America (Quantum Batteries), Quantum should not be understood *simply* as an EV or battery company. Instead, Quantum represents "the beginning of a different Bolivia, a new way of understanding what it means to be Bolivian."

SIX

———————

Green Energy Renegades and Lithium Futures

IN APRIL 2021, THE LITHIUM project in Bolivia was jolted by yet another surprising turn of events. Less than six months after the inauguration of Luis Arce as president and the return to democratic governance after the political and social crises of 2019–2020, Yacimientos de Litio Bolivianos, the state lithium company, organized a high-profile event in a ballroom at a hotel in the heart of La Paz's Zona Sur. The hall had been lavishly prepared for the guests by YLB. The state company, just emerging from the catastrophe of the year under the coup government of Jeanine Áñez and now humming with renewed momentum, had filled the spaces around the seating area and dais with different exhibits, which were meant to both attest to the progress made so far and provide a visual sense of the advances yet to come.

When Arce himself arrived, he was given a personal tour of these exhibits by a phalanx of YLB officials and technicians. As if to reinforce the serious-ness of the occasion, the YLB representatives from the industrial facilities at Llipi at the Salar de Uyuni appeared at the event in their industrial work clothes, including hard hats. As he passed by various displays, Arce paused the longest at one in particular: a small-scale model that showed what a fully built-out industrial complex will eventually look like according to the coun-try's lithium master plan, which, as we have seen, was itself in the process of being reformulated. Nevertheless, Arce carefully studied the miniature buildings, holding tanks, processing facilities, and warehouses—which had been fitted with tiny interior lights—with what was likely a mixture of anticipation and worry. By April 2021, the completion of the industrial lith-ium plant was still a distant prospect; indeed, it wouldn't be completed for over two and a half more years.

But the various realities—bitter and otherwise—of the actually existing lithium project in Bolivia were put aside for a remarkable four hours, during which time YLB announced an international public tender, or *convocatoria*. The convocatoria had become a governance mechanism that had been invested with much weight and significance in Bolivia, and not just in relation to lithium development. Beginning in the early years of Evo Morales's first presidency, a succession of Movement to Socialism (MAS) governments had emphasized the ways in which the recourse to open competitions for contracts with public entities showed how committed the MAS government was to the value of transparency, something that marked a stark contrast with the legacy of corruption and patronage that hung over a long line of past governments. The *practice* of transparency, however, as will be seen below, was another matter, but the launch of a major convocatoria like this one was nevertheless treated with a tremendous amount of solemnity.

So what was so important about the launch of this particular convocatoria that it warranted the presence of high-ranking government ministers, journalists, industry representatives (who had flown in from around the world), and the president of Bolivia himself? Unbeknownst to many, even within YLB, the Arce government had been working behind the scenes for almost a year, a process that began over six months *before* the October 2020 elections, while Arce was still a candidate running for office in the midst of both the coup and the Covid-19 pandemic. The Arce team had been actively exploring possible alternative technologies for producing lithium carbonate—technologies that didn't involve the highly inefficient and time-consuming use of evaporation ponds. Once elected, the Arce government expanded its investigations into alternative technologies and began exploratory discussions with companies and scientists from a number of different countries, a process that took place largely outside of public view. In other words, while the state lithium company attempted to restart the various operations at the salar, including both brine evaporation and the construction of the lithium carbonate plant, the Arce government was quietly taking steps to establish an entirely new pillar in the lithium master plan.

The April 2021 convocatoria announced this radically new dimension to a startled world—that is, the part of the world that had some stake or interest in Bolivia's lithium, myself included. Despite the fact that I was obviously deeply engaged as a researcher for almost two years by this point, like many in YLB, I hadn't learned of the background discussions and decision-making that had led to this moment, a closely guarded process that took place among a very small group of officials from the ministry of hydrocarbons and energy and the president's office.

By the time the Arce government was ready to go public with a decision that might, or might not, eventually transform not only Bolivia's lithium project but the fate of the green energy transition itself, it was clear—in ethnographic retrospect—that many smaller decisions had already been made, despite the fact that the convocatoria was presented as the *beginning* of a process of technological development and political deliberation. Even as the long-delayed construction of the industrial lithium plant continued in its course, Arce and a small group of ministers had become convinced that battery-grade lithium carbonate could be produced from brine by other means, through one of several methods of something called "Direct Lithium Extraction," or DLE.

Although I will examine the nuances and complexities of DLE in more detail below, here it is enough to say that DLE offers different techniques for extracting lithium ions from brine without having to wait for it to evaporate into a salt compound, which then, as we have seen, must be processed and chemically purified. As we would come to learn, the proponents of DLE claim that it has the potential to produce orders of magnitude more tons of lithium carbonate from the same quantity of brine, because the extraction process would have a "recovery rate" of upwards of 90 percent or more, meaning that almost no lithium from a given unit of brine would be lost, as it is now, to the uncertainties of solar evaporation and processing over many months under harsh environmental conditions.

After a long speech by the economist and technocrat Arce, in which he argued that it is necessary for Bolivia to pioneer DLE technology as a way to rapidly advance the country's infamously stalled lithium project, he turned the event over to the head of YLB, the ill-fated Marcelo Gonzales. The mild-mannered professor of physics from Potosí, who had only recently been appointed to the post, told me later that much of the background negotiations and decision-making around DLE had either taken place before he was appointed in January 2021 or had continued outside the direct purview of the state lithium company, which was expected to focus on its existing operations.

It was perfectly understandable, therefore, that he looked distinctly uncomfortable as he nervously read his way through a series of PowerPoint slides that gave information about the technical aspects of DLE; the government's plan to produce lithium carbonate at an industrial scale through DLE *by 2025* at the same time as the not-yet-completed industrial plant was also producing lithium from the evaporation ponds, *also by 2025*; and a number of companies that would be part of the convocatoria, including those whose representatives had decided to appear in person to make their own presentations. When he finished, Gonzales, who, as we have seen (in chapter 3), would be removed from office only three months later, turned the floor over to these company representatives.

Only six companies made their pitch at the convocatoria. Ganfeng Lithium, the $25 billion Chinese company, which is the third largest lithium producer in the world, concluded its presentation with the somewhat anodyne slogan, "Utilizing limited lithium resources to create a green, clean and healthy life for human development and progress." A representative from another major Chinese company, TBEA Group, extolled the fact that it had partnered with the Qinghai Institute of Salt Flats, a state research center that apparently has been conducting studies on brine and salar geology since the 1960s. The Russian company's presentation was dense with technical information, at least some of which had nothing to do with lithium but was about its innovations in nuclear power. As it turned out, the company, Uranium One, was a new subsidiary of the Russian state nuclear agency, Rosatom, which controls almost 80 percent of the world's nuclear power plant construction portfolio.

Stunningly, given the long history of animosity between the MAS government, especially under Evo Morales, and the United States, two US companies had been invited to make presentations. The first was given virtually, and in English, by the CEO of a company called Lilac Solutions, which I later learned was a lithium technology start-up backed by Breakthrough Energy, the "climate solutions" investment fund founded by Bill Gates in 2015. But it was the presentation by the second US company that drew my close attention.

The moderator invited the representative of a company called EnergyX to the podium. A surprisingly young man strode confidently to the microphone, where a MacBook was waiting for him, its gray metallic cover festooned with an EnergyX decal. He looked to be in his early thirties, certainly much younger than the representatives of the Chinese and Russian companies. He was also wearing what looked to my admittedly sartorially challenged eye like a designer suit, which he matched with just a trace of designer stubble on his face. He was introduced as Teague Egan, the company's CEO.

The slickly produced presentation—also in English—introduced the key members of the company, its main investors, and its DLE technology, which had been given a name followed by "TM," meaning that it had apparently already been trademarked (the name, not the technology). Egan explained that EnergyX was based in Austin, Texas, and then mentioned that the company was in the process of building a twenty-four-thousand-square-foot "innovation lab" right next to Tesla's new Gigafactory in Austin. A series of photographs, with Egan himself posing at different YLB facilities at the salar, was captioned, "Studying Uyuni for 3 Years." When he got to a slide entitled "EnergyX by the numbers," it claimed that the company had a market capitalization of $306 million. Egan paused to acknowledge the fact that the Chinese and Russian companies were much bigger—indeed, *very* much bigger.

Unfazed, however, he arrived at a startling slide, his *pièce de résistance*: a timeline of imagined lithium production in Bolivia if EnergyX were awarded the contract. The timeline showed five stages from 2021 to 2028: Pilot Plant, Demo Plant, Full Scale Stage 1, Commercial Stage 2, and Commercial Stage 3. Despite the fact that, as will be seen below, DLE encompassed a completely unproven set of technologies at industrial scale, EnergyX's timeline showed that Bolivia could be producing 45,000 tons of lithium carbonate by 2023, 150,000 tons by 2025, and an eye-popping *500,000 tons* by 2027. To understand how wildly ambitious—or, perhaps, just wild—this projection appeared at the time, the total amount of lithium carbonate produced in the entire world in 2021 was about 540,000 tons.

Egan's presentation was a bravura performance but also puzzling in so many ways. Who was this small upstart company from Austin that was claiming its DLE technology was going to allow Bolivia to transform the global economy by flooding the market with enough lithium to power *several* EV revolutions? And, perhaps more immediately, even personally: if the CEO of EnergyX had been carefully studying Uyuni for three years, why had I not heard of him before?

This chapter examines my various attempts to answer these and other questions, an ethnographic odyssey that ranged far beyond Bolivia itself. As will be seen, within the wider saga of Bolivia's lithium project and its fraught relation to the broader green energy transition, the story of EnergyX became a saga in its own right, one filled with both intrigue at the highest levels of government in Bolivia and with technonationalist hubris. This subplot to the wider lithium process was also marked by clashing approaches to innovation, risk, and even to capitalism itself.

The enlargement of the lithium project around DLE is certainly the most important shift that took place under the Arce administration, a shift that is part of a more general political economic realignment that I have described as flexible extractivism (see chapter 3). Given how much the global interest in Bolivia's lithium has intensified along with the international lithium rush, the opening of Bolivia's salares to what amounts to technological experimentation by foreign companies reflects several important dynamics.[1] First, in keeping with both the pragmatics of flexible extractivism and the deep concern by the Arce government to limit the amount of future public debt associated with the lithium project, the DLE initiative offers a way to pass the proof-of-concept costs on to foreign companies in exchange for more or less unfettered access to Bolivia's lithium reserves.

Second, in pursuing this course, the MAS government under Arce has all but acknowledged that the *ideological* dimensions of the initial lithium master plan no longer apply. If the original ambition was for Bolivia to vertically integrate the entire lithium "productive chain" as a definitive expression of a decolonizing "productive sovereignty," then the opening of the salares for DLE experimentation under the control of Chinese, Russian, or even American companies means that this grand ambition has already become a historic artifact.

And finally, it must be emphasized that the DLE gambit is not meant to replace the existing *productive* dimensions of the lithium master plan. The idea is that the evaporation ponds and industrial lithium plant will, eventually, begin producing lithium carbonate at scale, while the parallel DLE facilities are taking shape. But many technological and engineering questions continue to swirl around DLE at this point, even if it has been used to produce lithium carbonate in research laboratories and—reputedly—at a pilot level in Argentina and China. What would it mean to build an *industrial* facility based on these chemically intensive extraction processes, let alone a DLE plant that could produce five hundred thousand tons of lithium carbonate per year?[2] And if these technological and engineering problems cannot, in the end, be solved, what then?[3]

In using the trials and tribulations of EnergyX to reflect more generally on these wider stakes, the chapter also reveals something critical about the diverging moral, cultural, and political economic currents at the heart of the global energy transition. In particular, the ethnography of EnergyX, its vision, its business model, and its approach to technological development shines a light on the role of entrepreneurial capitalism. As will be seen, the ideological contours and praxis of entrepreneurial capitalism are often in significant tension with the political demands and social realities of resource-producing states like Bolivia.

In her study of energy entrepreneurs in Colorado, Mette High has shown how a culture of risk-taking by the lone and even heroic wildcatter, or energy renegade, is valued well beyond whatever can be extracted and commodified through the wildcatter's labors.[4] But as I found, an approach anchored in the values of the wildcatter clashes with expectations for things like risk-sharing, financial stability, and long-term commitment, especially when a historically vulnerable state actor is involved.

To be clear, this is not to say that the broader modalities of capitalism themselves necessarily clash with the needs and expectations of states in the midst of negotiating over the future of their natural resources, whether destined for green technologies or otherwise. Rather, it is to argue that under these particular circumstances, the circumstances that characterized EnergyX's presence at the center of Bolivia's big bet on DLE, entrepreneurial capitalism proved to be an especially bad, even jarring, fit. What I would come to learn, after following the story of EnergyX and the politics of DLE through many twists and turns, from Bolivia to Austin, Texas, is that entrepreneurial capitalism—and most notably its American variants—can only partly be understood as a series of business practices and strategies. Instead, through my research with an indelible exemplar of the category, I realized that entrepreneurial capitalism must also be understood as a form of self-expression, one whose real meaning and implications actually have little to do with problems of venture capital, innovation, or research and development.

In the next section, I continue the narrative of Bolivia's historic wager on DLE from Austin, where EnergyX has established a number of operations, including its corporate headquarters and a design facility in which they built a pilot "container" that would be used as a testing site in Bolivia during the first year of the convocatoria. This also gives me the chance to introduce the main players at EnergyX, including engineers, synthetic chemists, and, especially, Teague Egan himself. The chapter then examines DLE technology as an anthropological object, still from the point of view of people at EnergyX who are struggling to master the technology under a number of economic, political, and especially material constraints. As will be seen, one of the intriguing challenges for these extraction technologies is not the fact that lithium, as an element, is too complex; rather, it is that it is too simple.

I then return to the chronology of the convocatoria as it was experienced in both Austin and in Bolivia. This is where the clash between entrepreneurial capitalism and the imperatives of decision-making over the world's largest reserves of lithium by Bolivia's MAS government comes into full focus. If entrepreneurial capitalism is indeed as much a form of self-expression as it is a variant of the wider political economy, it is a form of self-expression marked by a self-assurance that can easily slip into hubris. I follow the trajectory of EnergyX and its maneuvering for the DLE contract in Bolivia to its conclusion, which has more than a few elements of a Greek tragedy.

The chapter ends with a short section that considers the wider meaning of this fraught episode for the future course of the energy transition. Much of the story is highly contextual; as we have seen throughout the book, Bolivia's unfolding lithium project is shaped by the country's history of resource conflicts, geopolitical entanglements, and social activism. Against this background, it is not surprising that the DLE process, and EnergyX's place within it, must likewise be understood through these contextual frames. At the same time, it is worth thinking through the context to broader questions about the status of entrepreneurial capitalism, especially now that major private entities like Bill Gates's Breakthrough Energy are starting to take the place of governments in the ongoing financing of the energy transition. The story of EnergyX and the turn toward DLE in Bolivia offers a critical lens through which to evaluate this broader, and arguably troubling, shift.

POWERING THE FUTURE FROM NORTHEAST AUSTIN

In his sprawling essay on what he describes as the "astonishing transformation" of his adopted hometown, Austin, Texas, the *New Yorker* writer Lawrence Wright paints a nostalgic picture of what the state capital was like when he moved there in the early 1980s: "Life in Austin was offbeat, affordable, spontaneous, blithe, and slyly amused, as if we were in on some hilarious secret the rest of the world was unaware of. Even then, the place had a reputation for being cool, but in my experience it was just extremely relaxed, almost to the point of stupor."[5] The city's unofficial motto was "Keep Austin Weird." Almost 20 percent of the city's residents in those days were students, faculty, and staff at the University of Texas. Wright's favorite restaurant was a "greasy spoon" called the Raw Deal, where the only choice was a steak or pork chops, which he used to pair with red beans and a can of cheap beer. A sign above the cash register reminded customers, "Remember: you came looking for the Raw Deal—the Raw Deal didn't come looking for you."

The astonishing transformation of this artsy, languorous, university, and political town also began in the 1980s, when the Reagan Administration decided it needed a national high-tech manufacturing center that could conduct research and produce computers and semiconductors to rival those made in Japan. In the open competition, fifty-seven cities submitted bids. To the shock of everyone, Austin was selected to host what was called the Microelectronics and Computer Consortium. At the same time, a drop-out from the University of Texas named Michael Dell started a computer company in Austin with $1,000 and a team made up of "three guys with screwdrivers."[6] By the end of the 1990s, Dell Computer was one of the largest personal computer manufacturers in the world and one of the largest employers in Austin, one with close ties to the university.

Twenty-five years later, Austin is now the "fastest-growing major metro area in America," a "turbocharged tech megalopolis [that is] being shaped by exiles from places like Silicon Valley."[7] One of these exiles is Elon Musk, who "has made Austin the centerpiece of his new Texas empire. In addition to the Gigafactory Texas . . . Musk's other businesses in and around Austin include the tunnel-drilling Boring Company; Neuralink, which is working on a computer-brain interface; and SpaceX, which is seeking to colonize Mars."[8] In 2018, another exile of sorts, Teague Egan, also made Austin his home. He came for many of the same reasons: to be part of the "turbocharged tech" milieu, to be close to the University of Texas, and to launch his own company—EnergyX.

One of the many jaw-dropping claims in Egan's presentation at the launch of the convocatoria in La Paz was that EnergyX was building a massive lithium "innovation lab" in Austin, but I found their existing footprint in the tech megalopolis decidedly more modest. Like most American cities, Austin is intersected by interstate highways, city roads, and various local thoroughfares. Outside of the city center, the public transportation system quickly thins out, making cars the primary means of travel. Only the foolhardy would think of cycling or walking across the vast and infrastructurally hostile distances needed to get from one part of Austin to another. To reach EnergyX, therefore, I was compelled to rent a (nonelectric) car.[9]

I headed north from central Austin on the ubiquitous Interstate 35, or "I-35," one of the major US interstate routes built during the Eisenhower Administration. If I had wanted to, I could have kept driving for another thousand miles, stopping only for fuel, all the way to Minneapolis, Minnesota. Instead, I turned off I-35 and continued east before heading into a district that was dotted with nondescript, prefabricated office parks and warehouses. Following the GPS, I made a series of turns and then, finally,

FIGURE 24. Small office complex in northeast Austin, 2022. Photo by author.

more than a year after the April 2021 convocatoria in La Paz, I arrived with great ethnographic anticipation at the homebase of EnergyX.

The immediate area was somewhat barren: monotone office buildings; wide streets without sidewalks (who would want to walk around here?); and empty fields of grass, which had already turned brown from the Texas sun even though summer had just begun. Following the directions, I pulled into the parking lot of a small, one-story office building that could have passed for a suite of dental offices anywhere in the suburban US. Across the street was the headquarters of Catholic Charities of Central Texas. And next to it was a branch of the Texas Department of Criminal Justice.

A sign in front of the building showed that there were actually three tenants in the smallish facility: EnergyX (listed on top); DiFusion Technologies, which designs "immunomodulatory" medical implants; and Water Lens, which makes water testing systems. As I approached the front door, I scanned the parking lot for EVs, something that gives me a provocative initial talking point in situations like this: your company is supposedly part of the transition to EVs, yet no one who works here actually drives an EV . . . However, I did see one: a Tesla S, which was the company's first model, with a license plate that read "ENERGYX" and was surrounded by a metal frame from the University of Southern California.[10]

Entering the EnergyX half of the building, the first thing I saw was an arresting wall-sized poster. It was a photograph of the Salar de Uyuni at twilight; the surface is already dark, and the sky is filled with the last wisps of a lavender sunset. The poster was entitled "HOW IT STARTED" and the text below narrates the official origin myth of EnergyX. And like all myths, it is partly true.

> EnergyX was conceived when Teague Egan was on a sabbatical exploring South America. As he ventured across the Salar de Uyuni, in the mountains of Bolivia, he came to learn this area was not only the world's largest salt flat at 4000 square miles, but also the world's single largest lithium reserve held in the salt brines below. Part of a larger area known as the "Lithium Triangle," up to 75% of the entire world's lithium is captured in this small area between Chile, Argentina and Bolivia. Primarily untapped, and with the biggest opportunity Teague had ever set his eyes on, EnergyX was born.

But to really understand what brought Teague Egan to Bolivia for the first time, we have to begin with his childhood and especially with his father, who provided Teague with extended lessons in entrepreneurial capitalism in the same way that some fathers in the US might toss the baseball around with their sons. Although he was born in Milwaukee, Wisconsin, to a family of Irish immigrants on his father's side who had come to the US during the potato famine, Michael Egan himself spent most of his early years in Florida. When he was only nine years old, he got his first job selling coconuts to tourists along US 1, the coastal highway that runs the entire length of the eastern side of the state.

After university, Michael Egan had a diverse career that included a stint in the US Army, a job designing food facilities for a university, and a post teaching at a hotel school. In the mid-1970s, he moved to Miami. During these same years, the US Congress was in the midst of a contentious debate over the airline industry. At the time the industry was tightly regulated by the federal government, which decided routes, fares, and most other aspects of air travel.

Given the bipartisan support to overhaul the system, Michael Egan envisioned a coming problem: if the airline industry was deregulated, the number of air travelers was going to rise dramatically. Deregulation would open the door for new airlines, more airlines would lead to more competition, and more competition between airlines would mean lower fares. How would these vast numbers of new travelers get around their destinations, especially since the US had invested in roads and highways and not public transportation?

As Teague explained to me, this was the moment when everything changed for his father and therefore himself. It was also the moment that gave rise to the single most important thing his father taught him: that business, especially one that required a heavy measure of entrepreneurial risk-taking, was *not* primarily about making money. Rather, it was about identifying a problem that for different reasons desperately required solving and then coming up with a solution earlier than others, one that, in retrospect, would appear obvious.

Michael Egan's solution to the problem of the coming waves of air travelers who would need a cheap and reliable mode of transportation was to buy Alamo Rent A Car in 1978, the same year the US Congress passed the Airline Deregulation Act. For the next twenty years, Egan built Alamo into one of the largest rental car companies in the world, focusing the business on the phenomenon he had predicted—the so-called leisure traveler. In 1996, when Teague was eight years old, his father sold Alamo to Wayne Huizenga, the owner of the National Football League's Miami Dolphins, for $625 million.

Teague himself grew up in Ft. Lauderdale surrounded by a considerable amount of wealth and privilege. He also began to see himself as an entrepreneurial problem solver from an early age. While he was still in high school, he proposed the first of what would become a series of audacious initiatives: a design for a retractable roof for Dolphin Stadium in Miami. As he explains,

> I was at a football game one time, and it started raining. And I noticed that probably three-quarters of the people left the game. And I started thinking about the lost revenues... So, I did some numbers, ran some numbers, and concluded that if it rains during three games a season, they are losing like $15 million or something like that. Right? I can't remember what the numbers were exactly. But how much would it cost to build a roof that would pay itself back in three years if it rains three times a year?

Fired up by both the apparent unimpeachability of the idea and the confidence that he, of all people, was the right person to take the project forward, sixteen-year-old Teague spent about a year working to bring it together. In partnership with his twenty-one-year-old cousin, Teague assembled a team that included an architect, an engineering firm, a company that "specialized in retraction," and a company "that made this special fabric, like a Teflon kind of fabric, that would allow light to come through but was obviously strong enough to stop the rain and allow airflow and things like that."

When the design proposal was finally ready, Teague used his ace in the hole: the fact that his father had not only sold Alamo to the owner of the Miami Dolphins but that the Egans and Huizengas were "family friends." The teenage Teague got his meeting with Huizenga, where he laid out his blueprints and explained his market analysis for the retractable roof. I asked Teague what happened, eager to learn the conclusion to the story.

> *Well . . . I had a few more meetings after that and then it never really went anywhere . . . Anyway, I graduated and moved to California, but he* [Wayne Huizenga] *sold the team the next year, and then the new owner built a roof the year after that.*

Using your design?

> *No, not using my design* [laughing].

At the University of Southern California (USC), Teague's entrepreneurial practicum continued. Much more than the content of his courses, he was constantly abuzz with new ideas, new problems that needed solving *by him*. During his sophomore year, while vacationing with friends on the tony island of Nantucket, he happened to stumble across two men engaging in a freestyle rap competition on the sand. Again, he was immediately seized with the idea of what should happen next: *All of a sudden, there's like a hundred people surrounding them as they are rapping back and forth. And I was like, this is amazing, I want to sign these two guys and start a record label.*

One of the two beach rappers wasn't interested, but the second was, and so Teague returned to Los Angeles and founded a record label called "1st Round Records" with only one artist. Teague brought him to LA several times over the next few months and they ended up making an album. When it was released in March 2010, it was an immediate hit, becoming the top downloaded album on iTunes for a period of time. As Teague explained, he had agreed on a 50-50 contract with the artist, in which they split all of the earnings after Teague had recouped his initial production costs. Teague said that they sold a million singles on iTunes and about 100,000 albums at $9 per album. The artist eventually left 1st Round to sign with RCA Records, leaving Teague with a lot of money and the desire for a new entrepreneurial venture.[11]

He surveyed the campus of USC and hit on yet another problem that needed solving, again, *by him*. The university is famous for its championship (American) football program, a program that has also been sanctioned many times over the years for violating the rules that govern the amateur status of what are referred to as "student-athletes." In those days, college athletes couldn't be paid for their labor and also couldn't receive any nonmonetary benefits or gifts from people outside the university—including sports agents—that might be construed as bribes. But Teague thought this created an opportunity: although sports agents obviously couldn't sign contracts with college athletes, why couldn't they form friendly relationships with them, so that when they *were* ready to enter the professional leagues, especially a multibillion-dollar league like the National Football League (NFL), they would be inclined to sign with an agent with whom they were already friends?

So during the summer between his junior and senior years at USC, Teague passed the test required by the NFL Players Association and became the youngest sports agent in the US. But his career as an agent would prove to be very short-lived. Because of the money he had made—"millions," as he put it—through his one-artist record company, he started to live lavishly, even for an undergraduate on USC's notoriously well-heeled campus. He bought a golf cart—his first experience with EVs—which he drove to classes and meetings, the rap from Boston's Boy blaring from the speakers. However, the good times came to a crashing halt when a freshman football player, the "number one recruit in the nation," flagged him down while he was driving to school. As he explains,

> *I see this star running back, he scored like seventy touchdowns in his senior year [in high school] in San Diego. And he sees me, "Yo, Teague," so I pull over. Then he says, "let me grab a ride to campus, I need to get to practice." "Yeah, sure, no problem." So, he hops in the golf cart. But the very next day, I see on the front page of ESPN* [a sport network in the US], *"USC running back suspended for taking extra benefit from agent."*

As it turns out, someone had seen Teague give the freshman football star a ride in the golf cart and had reported it to the university's athletic department, which then reported it to the national association that regulates university athletics. The repercussions were swift: Teague's recently issued agent's license was immediately revoked and USC notified him that they had begun a procedure to expel him from the university. In the end, he managed to negotiate a deal in which he agreed to have no further contact with athletes, a plea bargain of sorts that allowed him to graduate at the end of that year.

After university, Teague continued to live off the royalties from his record company while he tried his hand at a number of new ventures. But during this time, he also became fixated on the rise of Elon Musk as an entrepreneur. As he put it, he became an "Elon fanboy."

> I think he is widely recognized as one of the greatest entrepreneurs of our era, but to me of all time. You can put him in the same league with Thomas Edison and Leonardo da Vinci. But he has a different sense where he has the business mind too, in addition to the inventorship. Elon has mastered this thing and he has built things that have changed the world.

Teague took a half million dollars from the dwindling royalties from the record label and bought stock in Tesla when it was trading at $9 a share; he also bought one of the first Teslas, the same one I saw in the parking lot at EnergyX headquarters in Austin. Yet by late-2017, he was drifting, looking for a new problem that only he could solve. Not finding anything, he decided to organize a blowout trip to South America with "like fifty friends." The large group headed to Punta del Este in Uruguay for New Year's Eve, a seaside resort town known for its nightlife and "dystopian" exclusivity.[12]

After the multiday party was over, part of the group continued to travel through South America. Eventually, there were only two left: Teague and a friend. They decided to work their way down a "bucket list." They traveled to Machu Picchu in Peru, then went to the desert oasis village of Huacachina, in southwestern Peru, where they enjoyed "sandboarding in the dunes." The carefree adventure then continued:

> The [trip] then led us to Uyuni and we are on this tour and we are taking all of these Instagram pictures and all that . . . It was remarkable, one of the natural wonders of the world.

When you were out on the salar did you hear anything about the lithium project? Did you get close enough to see the facilities over in Llipi?

No, it was too far to see. But our tour guide starts telling us about how this is the world's largest lithium reserve. I didn't know the first thing about lithium. But when he said that, I thought, "I own a Tesla and Teslas use a lot of batteries." I've had my Tesla for five years. I could see this as a trend that was growing. I said to myself, "I made a shitload of money on Tesla, this is happening, I need to be a part of this." I didn't know how, but this is the world's largest salt flat or lithium reserve, and the Bolivians were essentially not producing lithium from it yet. This is a problem that I know I can solve. I mean, how hard can it be [laughing]?

Charged again with the buzz of entrepreneurial self-assurance, Teague returned to the US and got to work. He learned that lithium extraction through brine evaporation was inherently inefficient, so he wondered about other possibilities. Some reading about saltwater desalinization processes led him to contacts in the industry, who mentioned a professor at the University of Texas who was a leading researcher in membrane technology, which allows fluids to pass through a membrane while separating out particular elements. Teague then reached out to this researcher, Benny Freeman, a professor of chemical engineering, and explained that he wanted to know if any of the nanotechnologies he was working on could be applied to lithium extraction.

Freeman, who was actually conducting this research with a wider consortium of colleagues, including a team in Australia, offered to collaborate, but Teague would have to negotiate with the bureaucracy at the University of Texas that manages spin-off deals for technology produced by faculty members, the commercial rights for which often remain with the university. After months of back-and-forth, Teague and the University of Texas came to an agreement: they would license the membrane technology to him as long as Teague agreed to a number of initial and longer-term conditions. With the membrane technology now in hand, Teague was ready to take the next step: launching EnergyX. He settled on a phrase for the company motto that channeled a youthful audacity: "Powering the Future."

ATOMIC ORBITALS OF CONFOUNDING SIMPLICITY

When EnergyX started out in late 2018, the new company could only afford to hire a few researchers, most of whom were recommended by Benny Freeman. Then the Covid-19 pandemic paralyzed the new start-up's operations for much of 2020, although some laboratory work was able to continue. However, once work on the DLE technology resumed in earnest,

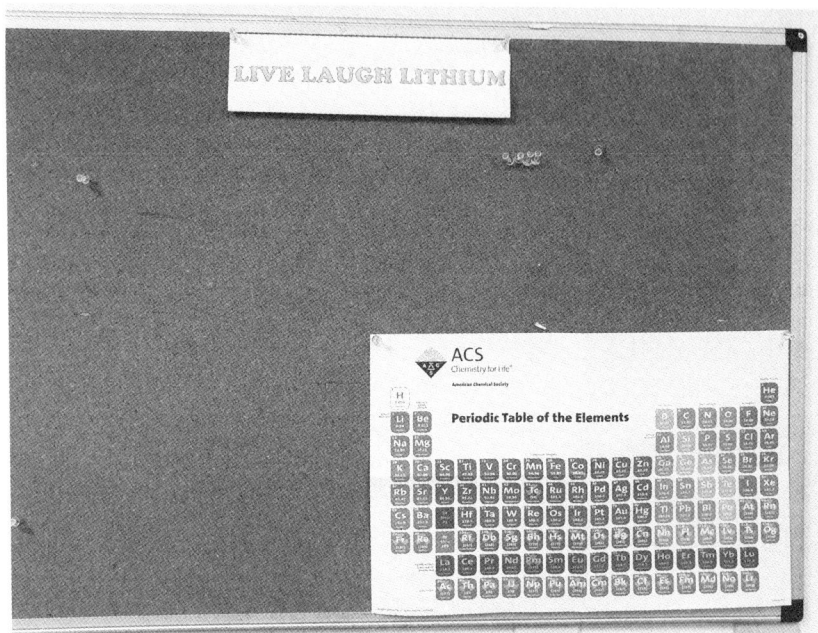

FIGURE 25. "Live Laugh Lithium," EnergyX offices, Austin, 2022. Photo by author.

the company began to realize that the membrane technology, no matter how it was formulated, confronted a basic structural problem. As other companies that are trying to develop DLE have also discovered, there is, and will likely always be, a tension between two material variables: selectivity and flux.

The first refers to the extent to which the membrane is designed so that it selects *only* lithium particles; and the second refers to the rate at which the brine passes through the membrane. To achieve higher rates of selectivity, the brine must pass very slowly, drip by drip. But such extremely low rates of flux, while acceptable in the laboratory, are not well-suited for industrial production, which would require exponentially greater volumes of brine to pass through hundreds if not thousands of membranes at exponentially higher speeds.

But membranes are not the only DLE technology being developed by companies, including by EnergyX, which eventually expanded into other processes. As Marcelo Gonzales, the former head of YLB, explained at the April 2021 convocatoria in La Paz, there are actually four distinct technologies for DLE: membranes; chemical absorption; ionic exchange; and solvent extraction, in which synthetic molecules are created that bond with lithium, and only lithium. Again, these four technologies, even then, were merely experimental designs, none of which had ever been tested beyond laboratory or pilot levels—that is, none had been used to extract lithium from brine at commercial scale.

And that is the crux of the problem for DLE, despite the ways in which it is being promoted as a historical turning point for the production of lithium carbonate. All of these technologies, in one form or another, have been used for years to extract elements. For example, membranes are now widely used in desalinization plants to produce fresh water from saltwater, often using a process of reverse osmosis. Solvent extraction is a chemically intensive and highly toxic procedure that is used by the world's largest mining companies to extract precious metals from mine tailings. This is a form of what might be described as *hyperextractivism*, in which companies seek a "recovery rate" as close to 100 percent as possible, despite the environmental, social, and economic costs.

But the application of these different technologies to lithium extraction from brine—which would likely involve similar levels of environmental damage and toxicity if and when they are implemented at industrial levels—must first overcome a dilemma with the structure of matter itself: the paradoxical fact that lithium is not too complex as an element to be extracted, but too simple. The technical aspects of this paradox were described to me by Jack Bender, who had worked for many years designing "extractants" for Germany's BASF, the world's largest chemical company. When I met Bender, EnergyX's chief synthetic chemist, he was in his mid-fifties and grappling with problems of a very particular kind.

Bender had grown up mostly in Europe, where his father was stationed at different overseas US military bases. Although his parents "were more on the artistic side, more creative," Bender realized from an early age that he had a special aptitude for math and science, especially chemistry. He returned to the US to finish high school and the rest of his education, which included a PhD in chemistry from the Colorado School of Mines, where he focused on molecular geometry. As a synthetic chemist for BASF, his job was to design molecules that would form the basis of "chemistries": chemical solutions that could be used to bond with the molecules of whatever element a mining company wanted to extract—gold, silver, copper, zinc, and so on.

BASF became the world's leading supplier of these extraction chemistries, which are used by all the usual suspects from the rogues gallery of major mining companies, including BHP, Rio Tinto, Glencore, and Anglo

American. But because these chemistries, with their synthetic molecules, are now well-established, there is much less demand for the services of synthetic chemists like Jack Bender. As he told me, eventually all the old-timers retired, leaving him the last synthetic chemist in the company. There wasn't any need to design new molecules for new chemistries because there wasn't a "critical" mineral that wasn't already being extracted—that is, until lithium, which is what brought Bender to EnergyX.

Bender tried to explain the paradox of lithium using a number of metaphors, before dipping into molecular theory when he could see by the somewhat blank look on my face that I wasn't grasping the problem. He said that solvent extraction—his specialty—worked like a lock and key. The given was the key, in this case lithium. As he put it, the key never changes. The job of the synthetic chemist is to design a lock that will work with the lithium key, and *only* with the lithium key, not also with the magnesium key, the potassium key, or the sodium key (all elements that are also abundant in Bolivia's brines).

To make matters even more complicated, the synthetic molecule, the lock, has to be designed so that it releases the lithium key during what Bender described as the "back end of the process":

> I haven't explained the back end of the process yet. So now I've extracted the lithium, it's in the [lock]. I want it back. Can I have it back please? So then, I go to the back end, and I turn it off. I turn it on at the front end, and I turn it off at the back end, and I know how to do that. But to turn it on and to turn it off, the [lock] will have to change shape, so you have to know how to do that too.

Moving on from keys and locks, Bender then turned to the structure of the lithium atom to try and explain why its sheer simplicity posed a formidable obstacle to DLE and therefore to Bolivia's momentous gamble on the technology. The problem comes down to "orbitals of bonding."

> In chemistry, we look at the orbitals of bonding, basically the valence of a molecule. For example, take hydrogen. It has one circular area that an electron can be in, super simple, crazy simple, there's nothing you can do with it, the electron is right there. As you go down the periodic table, you have all of these different orbitals, what we call p-orbitals, and d-orbitals, and f-orbitals, and s-orbitals, which tell you the way in which the electrons can be displaced around something like copper, which is much easier to do.

But why is lithium so different than copper?

The problem is that lithium has the smallest orbital. It only has two layers. One is completely full and can't be messed with. And the other layer has only one electron and wants to give it up, because it would much rather have no electron, and that's why you get lithium plus, that's ... why you have to kind of force it to be a metal ... And that's the complexity, that's why lithium is probably the hardest metal to go after, because it's too simple, it's the opposite of what you would think.

I don't understand. Doesn't a simpler key just mean you need to design a simpler lock?

No, actually, instead of being too complex to bond to, the more complex the orbitals are, the easier it is, because you can always say, "I know where I have to put the gripping parts, I know where I have to make the connections." But with lithium, it connects anywhere. So, the molecule has to be developed in such a way that it has just the right size pocket, with not too much charge, or not too little. It's like the three little bears ... no, not the three little bears, the ...

You mean Goldilocks and the Three Bears?

Yeah, Goldilocks and the Three Bears. The molecule is going to have to be not too hot and not too cold. It's going to have to be not too big and not too small, just right.

ETHNOGRAPHY WITH THE FALL OF ICARUS

In Breughel's *Icarus*, for instance: how everything turns away
Quite leisurely from the disaster; the ploughman may
Have heard the splash, the forsaken cry,
But for him it was not an important failure; the sun shone
As it had to on the white legs disappearing into the green
Water; and the expensive delicate ship that must have seen
Something amazing, a boy falling out of the sky,
had somewhere to get to and sailed calmly on.

FROM "MUSÉE DES BEAUX ARTS" BY W. H. AUDEN (1937)

While the scientific personnel at EnergyX struggled to manage the enormous technical challenges of DLE, events were unfolding far from Austin that would profoundly alter the company's fortunes. In the midst of a public health crisis and deepening political turmoil in Bolivia, the coup government under Jeanine Áñez finally capitulated to mounting national and international pressure and agreed to schedule elections for October 2020. In the months leading up to the vote, the MAS ticket of Luis Arce and David Choquehuanca gained strength in the national polls with each passing week, leading to a cautious sense of optimism among many in Bolivia that Arce would be elected and MAS's wider "process of change" soon restored.

During the campaign, Arce received advice and counsel from a number of people, in both official and unofficial capacities. Given the growing inevitability around his election as the October date approached, Arce's team started to formulate more concrete policy, especially economic policy. During the year of the coup and pandemic, the Bolivian economy—like many national economies—had contracted, leaving the economist Arce and his advisors particularly worried about which steps to take as part of a planned economic recovery in the new year. At the same time, the lithium project had clearly stalled in different ways and for different reasons during the preceding year, even if the full extent of the damage to the evaporation ponds and production facilities—much of it intentional—wouldn't be discovered until later (see chapter 3).

One of Arce's key unofficial advisors during this period was a Bolivian-American political scientist at Texas A & M University named Diego von Vacano. Von Vacano's father was Arturo von Vacano, the grandson of German immigrants to Bolivia who became one of the country's most prominent novelists and journalists during the 1970s. Although he himself was not a particularly political writer, Arturo von Vacano's circle of friends included some of the most strident critics of political repression in Bolivia, including Luís Espinal Camps, the Jesuit priest and human rights activist who was murdered by a death squad in 1980 in the early period of the Luis García Meza dictatorship. In this same year, Amnesty International intervened on behalf of Arturo von Vacano and his family and offered to resettle them in the United States.

The von Vacano family lived for a time in New York City, where Arturo worked for United Press International (UPI) as an editor and translator. When the offices of UPI were shifted from New York City to Washington, DC, the family moved to Maryland and then to the city of Alexandria, in northern Virginia. The family never returned permanently to Bolivia. Diego grew up in the US, attending three universities, including Princeton, where he earned a PhD in political science. As his academic career developed, he maintained contacts with friends and colleagues in Bolivian universities and in government, especially after Evo Morales came to power in 2006. He also became a widely consulted expert on Bolivian affairs in the international media. Because of his background, he was asked to provide advice on different issues by the increasingly confident Arce campaign.

As he considered different policy strategies, he wondered about lithium. As he told me, at the time, he had "no idea" about lithium. He knew very little about lithium as an element or about its status as a critical resource at the center of the global energy transition. He also hadn't been following the fraught efforts of the country to industrialize its own lithium reserves very closely, since his expertise was in political history and international relations. So, he did the most logical thing: he turned to Google. As he put it, "I literally Googled something like 'Bolivia' and 'lithium' because I really had no idea." He found a number of websites and YouTube videos that discussed different problems with traditional brine evaporation and these then led him to Benny Freeman from the University of Texas for the same reason Teague Egan had also found Freeman: because he was developing an alternative technology.

In what must have seemed at the time to Teague and EnergyX like an extraordinary stroke of good luck, something foreordained by the gods of entrepreneurial capitalism, von Vacano then reached out to Freeman and explained that he was an informal advisor to Luis Arce, who was running for—and would likely win—the presidency of Bolivia. As he said to Freeman, "Look, I'm also in Texas, it would be great to have an American company participate in Bolivia, not only in lithium but eventually in other ways, like supporting scholarships for Bolivian students that could be run through UT and A & M." Freeman then told von Vacano that as it turned out, he was *already* collaborating with just such a company, and it was also in Texas.

Even though Teague had founded EnergyX to "power the future" by solving the problem of "untapping" Bolivia's lithium, the company had no real plans for working with the Bolivian government or YLB. But then, completely out of the blue, Teague was contacted by Diego von Vacano on behalf of the likely next president of Bolivia. After a series of discussions, von Vacano offered to facilitate access to the main advisors in Arce's inner circle, including his oldest son, Marcelo Arce.[13]

Things started to move very quickly for Teague and EnergyX after that. He had a Zoom meeting with Marcelo Arce during which Arce expressed enthusiasm about the idea of working with EnergyX. Luis Arce won the election in October 2020, and Teague was invited to the inauguration ceremonies in La Paz the following month. During the visit, the president's son arranged a number of meetings for Teague, including with the newly appointed minister of hydrocarbons and energy, Franklin Molina, who was also enthusiastic about EnergyX's DLE technology. According to Teague, Molina told him, "This [technology] is great, let's put it in, let's get started." And then he met the new vice minister of advanced technologies, Álvaro Arnez, who would become the most vigorous advocate for EnergyX within the Bolivian government—at least for a time.

During this heady period for Teague and EnergyX, everything seemed to be going their way. It truly seemed to Teague like his salar epiphany, in which he saw himself as the entrepreneurial savior of Bolivia's lithium ambitions, was coming to pass in the most spectacular of ways.[14] A month after the April 2021 convocatoria in La Paz, EnergyX was notified that it was among eight companies chosen to move to the next stage. Each of them would be given brine samples from the Salar de Uyuni with which to test their different DLE technologies. They would have until May 2022, one year, to both demonstrate that their technologies worked and to prepare a detailed business plan in case they were selected.

Teague and EnergyX got to work almost immediately. Given the fact that the other companies—which included four separate Chinese companies and one Russian company (Uranium One, the subsidiary of Rosatom)—were significantly bigger and more heavily capitalized than EnergyX, it needed to find a way to distinguish its eventual proposal from the seven other bids. Since the Bolivian government planned to send brine samples to all of the companies, EnergyX considered approaches to testing that would give it an advantage over its competitors. And then it struck Teague: EnergyX would test its DLE technology *in situ*, on the wind-swept salar itself, under its unique weather and atmospheric conditions.

Using his close contacts at the highest levels in the ministry of hydrocarbons and energy, Teague managed to secure an extraordinary agreement: EnergyX would be allowed to install a "pilot plant" at Llipi within the gates of the YLB compound. But the plant had to be easily and quickly constructed or, better yet, shipped from Texas in prefabricated parts. The team at EnergyX soon realized that it was going to have to scale back its plans for the pilot plant. They settled on something more modest: a mobile laboratory inside of a small purpose-built container. In order to build and outfit the container in only a matter of weeks, Teague put the company's director of procurement and engineering in charge of the high-pressure rush job: Andrew Mullenax, a thirty-three-year-old idiosyncratic genius who had developed an international reputation while he was still in his twenties as someone who could, as he told me several times, build, fix, or design *anything*.

Mullenax and his brother grew up in Salt Lake City, Utah. His father had raced motorcycles ("dirt bikes") in the 1970s and worked as a janitor for a time before getting a job with a beverage company. He eventually worked his way up to a senior position and later became the executive director of the Utah State Fair, of all things. However, as Mullenax explained, his father was always building or repairing things: motorcycles, houses, John Deere tractors, cars. Andrew spent most of his free time with his father in the garage or on some building site, learning how to take machines apart, retrofit them with new parts, and solve problems of industrial design.

I asked him if he had always been interested in science and engineering in school and he said, "No, not at all actually." Nevertheless, as he put it,

> *It was always pretty easy for me. I always excelled. When I took those placement tests, it was always, "You are in the 1 percent of the 1 percent." When I was around ten or twelve, Bill Clinton wrote me a letter that said something like, "You win top 1 percent of 1 percent in the US for math or something."*

After high school, he enrolled at the University of Utah, where his major was, somewhat surprisingly, music theory. He was also an accomplished double bassist and performed with the Utah Youth Philharmonic. But a chemistry class awakened a desire to study science, and so he switched his major to materials science and engineering. Yet as he put it, he was the "worst student." Why?

> *Because I didn't do anything. I would try and bargain with the teachers so they wouldn't fail me. I would tell them, "Please just give me the book, I'll learn on my own." I told them, "You are not the guy who wrote the book, and you are teaching someone else's material." I just don't like learning from other people. I read about something and then I go and do it, and then I go to someone and ask very specific questions based on this experience, so I don't waste their time and they don't waste mine.*

Somehow, despite his unorthodox approach, Mullenax managed to graduate and enter the workforce. But he applied this same philosophy in his career. Even though he was in his mid-twenties, he didn't want to answer to supervisors and he didn't want power or money for himself. He just wanted to be left alone to figure out the most complicated engineering or logistics or procurement puzzles. Companies didn't really know what to do with him, so he became a kind of roving project engineer, someone who could be brought in to solve problems in a wide array of industries, from sodium cell technology to the guidance systems of Tomahawk missiles. As he put it, "I would get a call from someone on Christmas Eve, 'Hey, Andrew, we have a huge problem in Baghdad.' That's Baghdad, Iraq, not Bagdad, Arizona. I would tell him, 'Great, I'll go, I'll fix it.' I didn't need anyone else."

He eventually settled on gold mining, where one of his specialties became ore recovery. Gold mining companies sent him to mines around the world to recover gold, no matter the cost. According to Mullenax, this work took him to "every country in South America, North America, like 80 percent of Africa, most of Asia." I mentioned that YLB was getting a lithium recovery rate of 30 percent from the evaporation ponds at the salar and he just laughed derisively. *For me, the target is always 100 percent, all the time. I come from gold, which is so valuable that I'll build an entire second plant for $2 billion just to stretch it out of the tailings.* Really? *I've done it many times, absolutely. I'll buy a $200 million press, just to press the tailings.* That sounds bad for the environment. *Yeah, especially if we are using lead acid.*

When he arrived at EnergyX, Mullenax found the working conditions starkly different than what he had grown accustomed to. His workspace was a rented warehouse about a ten-minute drive from the company's headquarters in the same nondescript part of northeast Austin. Far from the unlimited budgets of the major gold mining companies he had worked for, Mullenax had to learn to design and build on the green energy start-up's tight initial financing. But he managed to get the mobile laboratory built in only three weeks, despite the "worst plastic procurement shortage" he'd ever seen. He also couldn't afford to buy the kinds of high-end equipment he was used to ordering, so he "took a bunch of pumps and took them all apart and retrofitted them with other parts."

In October 2021, the pilot container was ready to be shipped to Bolivia. Mullenax traveled to the salar region to be there when it arrived. He made quite an impression on the small group of Bolivians who were hired to help install the mobile laboratory and get it up and running. Mullenax was quite a bit younger than the Bolivian technicians; he didn't speak much Spanish; and his manner was informal in the extreme. People didn't believe at first that he was the chief engineer for an American energy and technology company. Mullenax ended up staying in the area for almost three months and only left when the experimental DLE technology was working to the point where it could be monitored by the local Bolivian team.

But then, in December 2021, as Teague and EnergyX were flying as close to the sun as they would ever get, at least in relation to Bolivia's lithium project, the beeswax in their wings began to melt. Rather than a sudden plunge into the sea, however, what happened next was more like a slow descent with the same ending. On December 16, an article appeared in the *New York Times* entitled, "Green-Energy Race Draws an American Underdog to Bolivia's Lithium."[15] The long-standing energy and business correspondent Clifford Krauss and the photojournalist Meridith Kohut had trailed Teague and other members of EnergyX during one of several lobbying visits the company made to Bolivia.

Krauss explained the stakes, "This nation of 12 million people potentially finds itself among the newly anointed winners in the global hunt for the raw materials needed to move the world away from oil, natural gas and coal in the fight against climate change."[16] And then the Houston-based energy reporter narrowed in on Teague himself, in a description that was certainly meant to convey admiration:

> Just as wildcatters have long sought riches prospecting for oil, the clean energy revolution is spawning a wave of gritty entrepreneurs who hope to ride a new boom, vaulting themselves into the intersection of geopolitics and climate change. Some are familiar names like Elon Musk at Tesla, while Mr. Egan and others are strivers looking for their first break in mineral-rich places like Bolivia, the Democratic Republic of Congo and the South Pacific.[17]

Beyond a short history of EnergyX and a brief biography of Teague, the article was marked by two main themes, both of which would have wing-wax-melting repercussions for the "gritty" entrepreneur. First, the article painted a striking picture of the improbable position of Teague and EnergyX. Teague "had never worked in the energy industry before starting EnergyX in 2018," a lack of experience that Krauss even suggests showed in his appearance, "[w]ith his hair slicked back, frequently unshaven and wearing his baseball cap backward." Krauss repeatedly makes the point: "Despite the bravado, he would appear to be an unlikely character to drive Bolivia's energy future. He has never worked in Latin America and speaks virtually no Spanish."[18]

But even more problematic for the fate of EnergyX in Bolivia, the article followed Teague as he engaged in what could be construed as backroom dealing with a number of key officials for the lithium project, including Carlos Ramos, who had just been appointed to lead YLB, and Álvaro Arnez, the vice minister of advanced technologies in the ministry of hydrocarbons and energy. In one particularly damning section of the article, Teague is apparently told by YLB officials that an "agreement approving EnergyX's project was virtually a done deal," after which the vice minister joined the EnergyX team "for a celebration at an elegant La Paz restaurant over plates of dried Amazonian catfish and roast pork with pear kimchi."[19] And this was all taking place while the convocatoria, a high-profile public tender that came with strict guidelines about fairness and transparency, was moving slowly forward.

Because the article was published in a US newspaper right before the end of the year, it got little immediate notice in Bolivia. But by the time offices reopened in early 2022, especially the offices of YLB and the ministry of hydrocarbons and energy, the impact of the article—which was instantly available in Spanish by clicking on "Leer en español" under the byline—was rippling through the corridors. The reaction was divided between incredulity and fear. People within both YLB and the ministry of hydrocarbons and energy were stunned to learn that a young, inexperienced, energy wildcatter from Texas apparently had the inside track to play a leading part in what could be one of the most important inflection points in Bolivian history, one with far-reaching implications for the "clean energy revolution."

But there was also a deep sense of unease. The article seemed to reveal high-ranking officials at the center of the lithium project operating well outside the boundaries of the MAS government's vaunted commitment to transparency. Moreover, dealings that included meals—and perhaps other inducements—at socially exclusive restaurants created an image that clashed with MAS's egalitarian and decolonizing ethos. People wondered: would those involved lose their jobs? It wasn't even clear whether or not Luis Arce was aware of the fact that government officials were giving their "blessing"— as the article put it—to EnergyX and its DLE technology. Perhaps the president's son, Marcelo, was actually pulling the strings? In a context in which a kind of informational "fog of war" prevails across the Bolivian government even in the best of circumstances, the *New York Times* article shrouded the DLE process with a particularly heavy aura of unsavory confusion.

However, the utterly predictable consequences within Bolivia of allowing a journalist and photographer to document his private meetings with high-ranking energy and lithium officials went largely unnoticed by Teague and the rest of the team in Austin. As Teague told me, he actually thought the article was a major boost for the company's international standing. Its publication had only increased his confidence that EnergyX would be the winner of the high-stakes public tender. But if the article showcased the company and its gritty entrepreneurial CEO, what about Bolivia, its government, its state lithium company? From Teague's perspective, it should also have been viewed not just positively, but as a historic moment for the country. Why? *I mean, it might not be, but this was probably the first time that Bolivia has ever been on the front page of the* New York Times. *This article is getting them global recognition for this amazing resource that they have, and I think it really painted them in a good light.*

With the May deadline for companies to submit their proposals and technological assessments to YLB fast approaching, EnergyX made another risky strategic decision. Perhaps still believing that the "blessing" of high-ranking government officials meant that the end result of the convocatoria was a foregone conclusion, the company decided to launch an Environmental, Social, and Governance (ESG) program in Bolivia. Teague hired two young Bolivians to organize and direct its ESG program, a private venture to invest social funds directly in communities in the salar region. One of the ESG directors was a graduate student at Texas A & M who had been recommended by Diego von Vacano. She then recommended a colleague who was a specialist in media and communications. Both of them had worked in the third Evo Morales administration (2015–2019).

They visited some communities and had begun to make inquiries through their old contacts in the government regarding the many potential legal and political hurdles to implementing a private development operation, but their first task was to write a script for a company publicity video. In early May, with less than two weeks remaining for EnergyX to finalize what would be for all of the companies an extremely complicated DLE proposal, the EnergyX team traveled to Bolivia to film the video.

The sixty-second spot features soaring music and rapid-fire scenes of the salar, Teague playing football with children, YLB trucks moving salt, Lake Titicaca, the EnergyX mobile laboratory, workers at the YLB industrial facilities, the EnergyX crew dancing joyfully in the middle of the salar, and scampering vicuñas. A female narrator solemnly intones (in Spanish):

The Bolivian people are full of energy. An energy that comes from Mother Earth, flowing from her volcanoes, mountains, and rivers. Bolivian energy nurtures the spirit of determined people, proud of their identity and culture. Bolivia has the necessary energy to transform the country, and the entire world. Bolivia has the power to be a leader in the global energy transition.

On this path of transformation, EnergyX accompanies the country, and the people. Sharing its sustainable technology, and creating local synergies to support health, education, and community. We will inspire the youth; we will build the future.[20]

As the narrator ends with "This is a new era, Bolivia's era," we see Teague standing on the surface of the salar with his back to the camera. He suddenly outstretches his arms and makes a grand gesture like a magician. The sky then fills with thousands of tiny particles, as if to say that he, and he alone, has the power to extract lithium through a sheer force of entrepreneurial will.

However, before Teague and EnergyX could accompany Bolivia on its path of transformation, and well before they could even *begin* to build the future together, disaster struck. Instead of a crowning moment of confirmation, a moment in which his deep sense of destiny was validated, things took quite a different, even bizarre, turn.

YLB had set a firm deadline for the receipt of submissions from the eight companies: 23:59 (UTC -4) on May 15. Teague had just returned from the filming trip to Bolivia and discovered that the task of pulling all of the different pieces of the EnergyX submission together was proving challenging. The results from months of testing in the on-site mobile laboratory were promising but the company submission had to include all aspects of the proposal, including a detailed plan for industrial production. Also, Teague had decided that the just-announced ESG program should be included in the submission, even though a number of components, especially agreements with communities, were missing. As he told me, "That is just red tape. I don't do red tape." And before it was submitted, the entire package needed to be translated into Spanish.

It had been decided that EnergyX's director of operations in Bolivia would translate the entire proposal. But in the last-minute rush—compounded by the start-up's small size and notable organizational looseness—Teague didn't send the full report to him *until May 15 itself*. Working as fast as he could for the entire day, the director of operations managed to return the translated version at 23:52, seven minutes before the deadline. Sitting on his living-room couch in his house in Austin, Teague now worked frantically to finalize the documents, which included having to compress a large PowerPoint file.

But as he scrambled at his computer, the deadline passed. At 00h02, Teague finally hit the send button, but the email immediately bounced. He stared at the screen, not knowing what had happened. As he put it, "I was literally sweating, shaking, sitting on the couch." And then he realized: he had used the wrong email address. Why? Because he had been using translation software to understand communications with YLB and had "translated" the YLB email address by mistake. He then entered the correct email address and finally managed to send the EnergyX proposal, ten minutes late.

The next day, May 16, in the presence of a notary public, YLB officially verified the receipt of submissions. Representatives of the companies were invited to be present, including EnergyX's director of operations in Bolivia, who flew in from Santa Cruz. YLB had received submissions from six of the companies in good order, including the four Chinese companies and the Russian company, Uranium One. The one Argentinian company, Tecpetrol, decided not to go forward with a proposal. And then there was the EnergyX submission: the email had been received, but it was obvious from the timestamp that it had been sent past the deadline. As EnergyX's director of operations in Bolivia told me, a distinct tension rippled through the gathering, which included the head of YLB, Carlos Ramos. What would happen to the EnergyX proposal? Could it even be considered at this point?

The answer came soon enough. After a couple weeks of internal deliberation between the president's office, the ministry of hydrocarbons and energy, and YLB, the state lithium company issued a press release on the status of the DLE process. It decided that it wasn't able to choose among the companies based on the preliminary results submitted in May, so it launched a new stage, in which the companies would be given the chance to demonstrate their technologies beyond the laboratory, at a pilot level. Six companies were chosen to move on to this next phase, the six that had submitted their proposals in May—that is, by the 23h59 deadline. EnergyX had been disqualified.

CONCLUSION: ENERGY TRANSITION AND THE LIMITS OF ENTREPRENEURIAL CONJURING

As I would later learn, the disqualification of EnergyX from the DLE competition in Bolivia on technical grounds, while perfectly understandable from a legal and institutional perspective, told only part of the story—and, as it turns out, not the most important part. As might be imagined, EnergyX wasn't the only company lobbying officials throughout this period; the Chinese and Russian companies were also actively promoting the advantages of their technologies and promises of long-term investment. One YLB official told me that the Chinese companies in particular were not happy with the implication that EnergyX was being privileged at different points in the evaluation process. And as we have seen throughout this book, Chinese companies have been centrally involved in Bolivia's lithium project from the beginning, a record of close and ongoing engagement that no Bolivian government under MAS would want to put in jeopardy. By quickly and publicly disqualifying EnergyX, the Bolivian government was able to restore faith in the DLE process, which had been shaken by the *New York Times* article, and assuage their politically and economically more important Chinese—and, increasingly, Russian—suitors at the same time.

But more generally, the unfolding of this episode and its aftermath revealed broader lessons about a kind of unscalability of a distinctive strand of entrepreneurial capitalism, especially within an energy transition that is supposed to be global in scope. If entrepreneurial capitalism is built around the myth of the individual business virtuoso, someone who seizes upon an important problem and then conjures a solution to the problem that no one else has imagined before, it also depends on a particular relationship with both risk and institutional financial stability. As my research with EnergyX in both Austin and in Bolivia taught me, the entrepreneurial start-up sells both an idea and a charismatic avatar of the idea; the product, or technology, that is meant to instantiate this idea comes later—or perhaps, in the end, doesn't come at all. But the goal of the entrepreneur is to keep this idea at the very core of the business while at the same time passing the risk of developing the business on to third parties for as long as possible—investors, customers, states.

Leaving aside the question of how important the entrepreneurial model has been to the history of capitalism itself, what concerns me here is the extent to which this model—and the myth of the unfettered renegade innovator—is particularly ill-suited to the social, political, ethical, and environmental demands of a multiscalar energy transition, especially in relation to "mineral-rich" states across the global South. Throughout the convocatoria, EnergyX had counted on the power of the idea that it could develop a DLE technology that was orders of magnitude superior to that of its competitors, an idea that its CEO promoted with an almost messianic "bravado," as the *New York Times* article put it. But as YLB officials involved in the DLE process explained, the country wasn't looking for an entrepreneurial savior, especially not from the United States. Although the Arce government had moved away from the more full-throated anti-US and anti-imperialist rhetoric of the Morales years, a strong anti-US current remained, especially among the MAS rank-and-file.

Even more problematic for EnergyX and the viability of the entrepreneurial model, however, was the belief that its strategy required YLB to assume much of the risk, both financial and technological. YLB officials understood a "start-up" (they used the English word) to be a new, small, and potentially fragile company with relatively little capital and financial stability. And EnergyX's DLE solution was also viewed with skepticism, not because of the technology itself, but because of what Teague had described at the April 2021 launch event in La Paz as its "unique approach." The company's technology was being designed to be implemented at later stages in the traditional brine evaporation process, while its competitors were proposing DLE technologies that would eventually replace brine evaporation altogether. EnergyX claimed that this approach was being developed to improve "efficiency." However, some officials in Bolivia—especially those with technical expertise—read this as a way of passing the burden on to YLB, which would be required to maintain the costly and inefficient system of pumping and evaporation.

Instead of taking on these various risks, the key policymakers at the center of the lithium project in Bolivia preferred a very different model, one in which the state lithium company—by necessity—partnered with well-established and deeply capitalized foreign companies, ideally those that were either state companies like YLB or had close backing from their respective states. More than one YLB official told me that Chinese companies, in particular, fit this preferred model, especially given that China was still officially a socialist state, like Bolivia. The belief was that, like Bolivia's state companies, Chinese enterprises worked for a higher purpose beyond mere generation of revenue.

This ideological interlinkage, as tenuous as it might be in practice, was something that was seen to transcend the normal transactional alignments of capitalist exchange. A renegade energy start-up might very well produce the most innovative DLE technology. But winning the contest for the most innovative means little to historically vulnerable, if mineral-rich, states like Bolivia—especially if it requires these states to take on the lion's share of the risk in what amounts to a subsidized proof-of-concept project, one targeted primarily to satisfy the interests of future investors in the game of multistage venture capital. The turn to DLE might very well be the shift that finally puts Bolivia—and the global energy transition—on a "path of transformation." But if so, it will almost certainly not be a path conjured into existence with a flourish of the entrepreneurial wand.

Conclusion

THINKING THROUGH BRINE

AS IT FLOWS UNSEEN within the prehistoric depths of the Salar's porous, tectonic, many-layered geology, the salty liquid at the center of the global energy transition carries with it more than lithium: it also carries lessons. In thinking through brine and through the other opaque materialities entangled with the futures promised by the transition from fossil fuels to lithium-powered forms of energy, we can understand much more clearly the ultimate folly of these futures, the ways in which the lithium imaginaries of the present are not actually projections of a tantalizing—if unrealizable—*alternative* future but rather projections of an all-too-realizable past and present *in the future.*

When brine is pumped up from its subsurface universe, one in which it remains protected from the elements by the reflective power of the salar's halite crust, it begins an agonizingly slow process of transformation, not just from brine into evaporated salt compounds but from mineral-rich brine into objects that create energy, use energy, and store energy. At each point, or node, in this vast lithium energy assemblage, this process of transformation is marked, above all, by resistance. The seemingly intractable imperative to overcome this resistance is the same imperative that drives the wider global political economy within which lithium-ion batteries and electric vehicles are supposed to provide a technological fix to the planetary condition of Anthropocenic deterioration. But what does it mean to privilege, even celebrate, the technological fix, the overcoming of "resistant materiality," over the illuminating fact of resistance itself?[1]

To return to brine: let us follow this path of resistance to its endpoint. As we have seen, the material complexities don't smooth out along the way; instead, they accumulate from one moment to the next, transferring their resistance, while adding social, political, historical, and moral strata. When

the many months of solar radiation have rendered brine into a heterogenous white powder, the limits of production will have already undermined the process: about 70 percent of all the lithium that was present in the first evaporation pond will have escaped. It eventually makes its way below the surface once more to rejoin a new liquid current on a new course through the salar's fractured underworld.

To get from the mountains of salt compounds that remain to another white powder—battery-grade lithium carbonate—requires hundreds of other small struggles with resistance, struggles that become more and more chemically and technologically intensive. Indeed, the fact that the difficulty in overcoming resistance to the process of purification becomes exponential with every tenth of a percent is a microcosm of the entire problem. How can it be that the laborious and toxic process for obtaining three extra tenths of a percent in purity of a salt compound that holds a nonrenewable resource—say, from 99.5% to 99.8%—is one of the foundations on which the "clean energy revolution" depends?

But the trajectory of resistance is only beginning. As we have seen, the lithium molecules trapped within this ultra-pure compound are destined for their own waystations on the longer path toward "sustainability." The ethnography of one small battery pilot facility in a rural valley outside of the city of Potosí revealed just how fraught the struggle is to synthesize cathode materials, to interpret and verify their crystalline structures, and then to heat them to very high temperatures—to calcinate them—without altering the extremely delicate chemical mix. And all of these struggles are needed to produce only one of the three main components of a lithium-ion battery, the cathode. The anode and electrolyte are produced by overcoming their own resistant materialities, even if these are less formidable.

Although these three components are the most important, battery assembly requires other lines of resistance to be overcome, including the provision of metal casings and internal electronics (each with their own supply and extractive histories), testing for battery integrity, and conforming with whatever design standards have been established. And the process of overcoming resistance that began in the dark depths of the Salar de Uyuni is just to produce a battery cell. And what is its capacity? What can it be used to power? The production of *high-capacity* battery cells, those that are used as part of the large battery banks that power EVs, involves the same sequence of steps, except that resistance is magnified at each step and along every axis. To produce battery cells destined for use in an EV with, say, a 220-kW engine, demands a much more intensive struggle against resistance than to produce a battery meant for a handheld laser pointer.

At the kinds of industrial scales that are needed to supply lithium-ion batteries for the global EV market, the number of cells needed are staggering—as are the resulting amounts of waste, toxicity, and unsustainable resource consumption. Depending on the model, a *single* EV can use up to *nine thousand* individual battery cells. The largest battery suppliers in the world produce tens of millions of battery cells per year. And although battery assembly at these levels is obviously automated, its trajectories of resistance likewise begin at the point of extraction, if not yet at the Salar de Uyuni, then at the Salar del Hombre Muerto in Argentina, or at the Salar de Atacama in Chile, or at the Greenbushes hard-rock mine in Western Australia.

From here, lithium molecules—which might have been trapped for millennia in brine but are now encased in one battery cell among thousands in a massive, weighty, battery bank—continue along an increasingly complex path of resistance. This path becomes in different measures economic, political, and social—in addition to material. It is here where lithium's resistant materialities become part of the labyrinthine matrix of struggles that are embodied in the replacement of fuel-burning cars and trucks with EVs. Automakers must overcome hundreds of resistance points to convert their production lines, resistance that involves supply chains, labor relations, automobile design, and company strategy.

And once EVs have been manufactured, who will buy and use them? The replacement of fuel-burning cars and trucks with EVs, a lodestar that guides the global energy transition, means exactly that: *replacement*, the complete exchange of one transportation technology for another. But as the short ethnographic video I shot in front of the Quantum showroom in Cochabamba suggests—a video that captures a quotidian moment of fossil-fuel burning tumult—resistance to the EV "revolution" is pervasive. Indeed, one might say that resistance to the global electrification of mobility is itself global—if not within energy policymaking, then within the everyday lives of energy, within the cultures and praxis of energy around the world.

But for those who *have* joined the revolution, the path of resistance continues, a path strewn with obstacles that must be overcome. The lithium ions captured within the thousands of battery cells in a single EV must return to the anode after the car has been used and the energy discharged, whether over thirty miles, like a little Quantum E4 (whose battery banks use far fewer cells), or over three hundred miles. Where will this recharging take place? Here, the wider trajectory of resistance, which began at the point of extraction, at the point at which brine was extracted from its subterranean geology, becomes infrastructural. As Carlos Soruco told me, beyond questions of cost and a kind of deep-seated technological skepticism, the greatest impediment to bringing the EV revolution to Bolivia is the utter absence of public charging stations outside of a few exclusive neighborhoods in Cochabamba, where several were installed by Quantum itself. The same story of infrastructural resistance can be told for almost every city, every highway, every village, every apartment block, and every parking lot in the world.

And even if an underlying extractive imperative has driven the process forward to this node in the wider lithium energy assemblage, the node at which lithium wrested from brine powers a carbon-free vehicle, at least in relation to tailpipe emissions, how is the charger that recharges the vehicle itself powered? Where does the electricity come from? What does it mean that with the exception of a handful of global outliers like Norway, whose national electricity grid is almost entirely dependent on renewable hydropower, most electricity is still produced by fossil fuel–powered turbines? At this stage, the long path of resistance, which began at the Salar de Uyuni—or at another site of extraction—begins to loop back on itself.

This is because the electricity needed to charge hundreds of millions of new EVs around the world, which would mark the kind of total vehicle replacement imagined by international energy policy, would require vastly greater amounts of electricity to be generated, which, under current conditions, would mean that the so-called EV revolution would further *exacerbate*—not mitigate—the climate crisis. And this is before we consider all of the adjacent harms and forms of toxicity that would accompany a lithium energy assemblage taken to its logical conclusion, including the depletion of fragile water tables around zones of extraction, the creation of untold amounts of industrial waste, and the problem of disposing of billions of spent battery cells (despite the hope that lithium might someday be recyclable, like the lead in lead-acid batteries).

In other words, to think through brine, to trace these trajectories of resistance into the futures imagined by the blueprint for a global energy transition, is to understand the degree to which "transition" means small shifts in form while preserving as much of the content of the present as possible. In a way, lithium is the ideal avatar for this troubling realization. Its moral valence is clearly different than that of oil, gas, coal, and other icons of both nonrenewability and our late-carbon age. But unlike solar, wind, or hydropower, lithium itself is not a renewable resource. It is a "critical" mineral, a metal (albeit an alkali metal), an element that is abundant on Earth but only in high enough concentrations in very few places to be extracted at scale.

And yet, despite it all, this particular nonrenewable resource is one of the linchpins of our response to global heating and its manifold associated crises. If it's not clear by now, it should be: the idea that we will be able to extract our way out of the climate crisis, which is only one among several of the gravest human-induced planetary ravages, should itself be resisted. It should be resisted, that is, if we are serious about a program for planetary recovery that seeks to undo the devastation of a centuries-long assault on the Earth's biosphere, an assault that began as the responsibility of the few but which has become more global over time.

If we *are* serious about this long-term program—which must be followed at the kinds of international scales that are impossible to imagine under current conditions—then the revolution will not involve the simple replacement of one kind of vehicle with another, both of which are being produced within the same capitalist system, animated by the same extractivist imperative, and shaped by the same relationships between resources and geopolitical power. As the feminist intellectual and poet Audre Lorde put it, the master's tools will never dismantle the master's house.[2] Instead, much more radical changes to sociopolitical, economic, and ecosystemic life will be needed. Cymene Howe and Dominic Boyer describe this other world: "This project will be utopian in the sense that it will have to make a world that has not yet existed. It will be revolutionary in the sense that it will not be accomplished by technology, or markets, or violence, or anthropocentrism, or any of the other behaviors and attitudes that brought us here in the first place."[3]

If we, collectively and globally, are not serious about undertaking the kinds of far-ranging alterations to the present that will be necessary, then there *will* still be a place for the extraction of lithium and the push to replace fuel-burning cars and trucks with EVs. This is because the burdens of "life in capitalist ruins" would certainly be eased if they were faced in a climate whose local atmospheres were marginally cleaner, even if global heating itself continued to worsen.[4] Nevertheless, an Anthropocenic world in which the main marker of transition is that people drive electric, instead of fuel-burning, vehicles, is not a new world. It is rather a world in which the long-term "signatures"—as paleontologists describe them—continue to point toward a future in which the "behaviors and attitudes that brought us here in the first place" have, if anything, deepened and become more ossified, more suggestive of approaching *dystopia*.

But where does all of this leave Bolivia and its fraught, ever-unfolding, ever-shifting lithium project? As we have seen, Bolivia's lithium ambitions appear very different when viewed through the micro-lenses of history, political conflict, and biography—institutional, social, and individual. The fact that the lithium project is formally under the supervision of the state doesn't change the ways in which the project is itself an assemblage of disparate practices, actors, and visions for the future. Even so, it is significant that lithium industrialization in Bolivia is being undertaken by a state that is committed—beyond mere political rhetoric—to a policy of social redistribution of public revenues earned through extraction, despite the many internal divisions and cultures of local opposition, especially along what Penelope Anthias has described as the country's "hydrocarbon-conservation frontier."[5]

Yet it would be wrong to demand of policymakers in Bolivia that they shape the future of the lithium project in response to these wider debates and critiques—that is, in response to the fact that, at a global level, resource extraction will never pave the way to a post-Anthropocenic world. Although some social and Indigenous movements within Bolivia have mobilized for non- or post-extractivist visions of development, the country itself shouldn't be held to account for its enduring economic dependence on raw materials. Climate *injustice* takes many forms, including the expectation that small mineral-rich countries like Bolivia, which were plundered for centuries, should leave their resources untouched in order to mitigate global harms for which they bear no responsibility. Instead, I have no doubt that Bolivia—at least under a MAS government—will continue to yield to the siren song of lithium, the messianic quest to "untap" the world's largest reserves of a resource at the center of an energy transition that seduces precisely because it is so elusive.

Afterword

AS I HAVE EXPERIENCED with most of the other books I have written, the process of writing and revision is frequently shaken by unanticipated events or developments, some of which involve the people, places, and institutions that figure in the manuscript, and some of which involve wider processes about which the book, ideally, has something important to say. The question is always: do these events or developments require a rethinking of the book's *fil rouge*, the central thread (or, perhaps, threads) that runs from the first chapter to the last? In the case of a work of ethnography, which functions as a critical snapshot of a discrete period of time—in this case, a period of four years—the fil rouge problem is perhaps less acute, given that this snapshot has already passed into history at the moment it is evoked in text.

With *Extracting the Future*, certain processes that were delayed or uncertain were finally resolved; as might be expected, different key figures in the book have experienced life changes, some happy, and some tragic; and the broader global market for lithium and EVs has undergone some changes. But the main outlines of the narrative remain very much intact, both within Bolivia and across the many nodes of the lithium energy assemblage within which Bolivia's lithium project remains as entangled as ever. Moreover, the main argument of the book—that an energy transition undertaken through the mechanisms of global capitalism and dependent on the extraction of nonrenewable resources like lithium will never seriously mitigate the ravages of the climate crisis—has, if anything, become more urgent in the months since the book was completed.

Nevertheless, while there is still time, it is important to commit some of these events and developments to writing, for ethical reasons if for nothing else. To begin with the wider frame of reference—the global lithium energy assemblage—the market for lithium and EVs has seen some shifts, most importantly in the price of lithium carbonate, which has experienced a relative plunge from the historic highs that shaped Bolivia's lithium project during the principal years examined in this book (2019–2023). This price fluctuation can be explained by the short-term oversupply of lithium and batteries against an equally short-term weakening in the global EV market, as the shift to EVs is proving slower than anticipated.

But the mid- and long-term picture, the one captured in this book, remains unchanged. The demand for lithium is still expected to *double* by 2030 and continue to grow exponentially in the years and decades thereafter, a demand that will never be met without access to Bolivia's likely thirty-million-plus tons. In other words, Bolivia's ever-unfolding lithium project will continue to remain at the center of this new resource geopolitics even as governments rise and fall, perhaps even as the era of MAS gives way to a political future for Bolivia that is impossible to predict at this point.

The critical importance of Bolivia's lithium to the current and future political economy of energy transition, one fed by extractivism and shaped by the logics of consumption, economic growth, and capitalist market relations, has not been significantly altered even as the global rush to find new lithium deposits has become more fevered, and, in certain instances, more violent. For example, a notable change in the status of global lithium reserves has included a marked increase in reserves "measured and indicated" in the United States. But with the coming turn in policy against so-called green energy technologies by the second Trump administration and the corresponding return to fossil fuel-centric policies by the US government, it is difficult to say whether or not the US will continue to remain far behind especially China in the struggle to control the most important nodes in the global lithium energy assemblage.

Yet even if the global rush to discover—and industrialize—new lithium deposits doesn't change the long-term importance of Bolivia's lithium reserves, this doesn't mean that this global rush isn't taking place without serious consequences. In at least one instance, I was myself swept up by the violent consequences of the lithium rush. In August 2023, I was contacted by a Serbian journalist working for one of the last remaining independent news outlets in the country, which has experienced restrictions on press freedoms and civil rights during the government of Aleksandar Vučić. The journalist had come all the way to Oxford to record and film a podcast with me about a controversial deal between Serbia, the EU, and the private mining company

Rio Tinto to develop a hard-rock lithium mine in the middle of the fertile Jadar River valley. Significant opposition to the lithium mine was building in Serbia and the journalist wanted to offer a perspective about the project that wasn't associated with one side or another in what had become a major national conflict.

Our lively discussion took up a range of topics, including the proposed project in the Jadar River valley, a discussion that was also informed by my research on lithium extraction in Bolivia and its fraught relation with energy transition and the climate crisis. Within days, different videos of the podcast had gone viral in Serbia, with over 100,000 views and thousands of comments across the different platforms. Using Google Translate, I followed these discussions for a few days but stopped after reading one too many anonymous progovernment comment that was abusive or threatening, mostly to the journalist who had interviewed me.

Days later, Belgrade was paralyzed by a massive protest against the lithium project; according to the journalist, our interview had helped fuel a national opposition to the mine. However, the protests not only failed to stop negotiations between the Serbian government, EU officials, and Rio Tinto, they provoked the Serbian government to move against the activists. Those protesting the lithium mine were labelled "ecological terrorists" in official state media and the government set up a website where citizens could "report an ecological terrorist." This website also included the names, photographs, and doctored biographies of some of the more notable figures associated with the protests, including the journalist who had interviewed me in Oxford.

Within Bolivia, there have also been a number of important developments, some equally shocking in their own way. In August 2023, Carlos Ramos was fired as the head of YLB, exactly one year after he had told me that he knew his days were numbered with the state lithium company. To recall, Ramos confronted a series of impossible construction and production targets as part of the government's general hydrocarbonization of the lithium project and its parallel turn toward DLE as a new technology. What was surprising wasn't Ramos's ouster but the announcement of his replacement: Karla Calderón, his thirty-two-year-old former right-hand advisor, who became the first woman to lead YLB in its history. As the ambitious Calderón put it, at the press conference announcing her appointment, she would do her utmost to make sure that YLB became Bolivia's "jewel."

Several months later, in December 2023, the industrial lithium carbonate plant was finally completed, nearly four years after the date agreed upon in the 2018 contract between YLB and the two Chinese contractors. But almost immediately, YLB officials realized that there were major problems with operations at the new plant, mostly to do with the preexisting evaporation ponds on which it relied. Although, as we have seen, the ponds suffered from both passive and active neglect during the coup year of 2019–2020, YLB officials were now claiming that the problems with the ponds began much earlier, during the years in which the lithium project was managed by GNRE under the control of COMIBOL, Bolivia's state mining company. These problems apparently included everything from corruption in the awarding of subcontracts to the use of faulty geomembranes in the waterproofing of the ponds themselves.

In a stunning development, YLB announced in April 2024 that it was filing criminal complaints against eleven named and unnamed former officials for financial crimes against the state, including breach of official obligations and economic damage to the state lithium company. The first official to be named and then arrested was Luis Alberto Echazú, the former head of GNRE. But Juan Carlos Montenegro learned that he was also under investigation. Montenegro, who was a regular source of important information for our project over the years and an engineer and lithium expert with an impeccable profile, had been the widely admired inaugural head of YLB when it was launched in 2017. But when I last spoke with him in 2023, after he had returned to his academic post at La Paz's *Universidad Mayor de San Andrés*, he had become increasingly pessimistic about the future of the lithium project.

What he couldn't predict, however, was the way his service to YLB and to the Bolivian state would become enmeshed in an increasingly bitter struggle between the government of Luis Arce and the supporters of former president Evo Morales, an internal struggle between rival factions within MAS that was becoming ever more heated as the August 2025 national elections approached. As the long-term lithium project in Bolivia continued to suffer delays, experience operational problems, and buckle under what I have described as the force of historical reckoning, there is no doubt that the Arce government saw the criminal complaints as an opportunity to kill two birds with one stone: first, they were able to identify high-profile scapegoats for the various problems at the heart of the lithium project, thereby deflecting attention away from questions about how these problems might have been exacerbated after Arce took power in 2020; and second, the criminal complaints served as a broader indictment of the different Morales governments, despite Arce's key role in these governments.

In any event, the looming legal and personal tsunami proved too much for the mild-mannered professor of engineering. On April 24, 2024, Montenegro was found dead of an apparent suicide. According to his lawyer, he left a one-page note of explanation, which was later published in the Bolivian media. In it, Montenegro denounces the Arce government's management of the lithium project, specifically its decision to turn toward DLE as an unproven and risky new technology for lithium extraction. Montenegro goes on to say that he refuses to be "shamed and humiliated by a rigged justice system that sells itself to political power or to the highest bidder." At the bottom of the carefully typed note, after asking for forgiveness from his loved ones for the pain his decision will bring them, Montenegro adds a final plea in blue handwriting, as if he wanted to leave no doubt: *Soy inocente*, I am innocent.

In the meantime, the DLE process in Bolivia had moved on considerably from the years of the convocatoria. Between January and June 2023, in the waning months of Carlos Ramos's tenure at the head of YLB, the government announced the results of the competition for the DLE contracts, results that came as no surprise given everything that had come before. Deals were made with the giant Chinese battery maker CATL, the Chinese consortium Citic Guoan Group, and the Russian state company Rosatom, for almost $1.4 billion in investments in DLE operations at all three of Bolivia's major salares: Uyuni, Coipasa, and Pastos Grandes. These deals were followed about a year and a half later with another major DLE contract, this time with the Chinese consortium CBC, worth another $1 billion. However, when this last deal was announced by YLB, it wasn't by Karla Calderón. Despite her background and complete commitment to the success of the lithium project, she lasted only about a year in office. In September 2024, she herself was replaced by Omar Alarcón, who had spent almost his entire career with YPFB, Bolivia's state oil and gas company. His appointment was perhaps the final step in the hydrocarbonization of lithium in Bolivia, the clearest sign yet that lithium had been reduced by policymakers to simply one more nonrenewable resource whose only value for Bolivia was as the basis for new state revenues.

And if the DLE process in Bolivia had moved on, what about EnergyX, the renegade American green energy start-up whose trials and tribulations are in part documented in chapter 6? As it turns out, my reading of this story as a classic Icarus narrative, of a self-confident young entrepreneur who flew too close to the sun only to crash into the sea, was incomplete at best. The years after May 2022—when the company was disqualified from the DLE convocatoria officially because it had missed the submission deadline by mere minutes but unofficially because of a range of factors that I analyze in the chapter—have in fact been very good for Teague Egan and EnergyX. Indeed, one can say that Egan quickly burst from the sea to soar even higher on the wings of his entrepreneurial ambition.

Just over two months after the May snafu, Global Emerging Markets, highlighting EnergyX's environmental, social, and governance (ESG) profile, announced a commitment to invest $450 million in the company. EnergyX then quickly pivoted from Bolivia to expand its DLE projects to other countries, including the United States. In April 2023, General Motors agreed to invest $50 million in EnergyX in order to "unlock" the US's lithium supply. And in October 2024, the company announced yet another major round of funding, this time from the digital platform DealMaker, which brought in an additional $75 million.

Finally, in November 2024, EnergyX inaugurated its new "innovation lab" and headquarters in Austin, Texas. At forty-thousand square feet, the new facility is almost twice as big as the planned innovation lab mentioned by Egan at the April 2021 event in La Paz that launched the DLE convocatoria. Although I am as convinced as ever that the mechanisms and logics of entrepreneurial capitalism will never pave the way toward the kind of radical transition that will be necessary to truly confront the consequences of our rapidly, if unequally, deteriorating planet, even I must admit that we live in Teague Egan's world. As he put it, on the day he inaugurated the company's new mega-headquarters:

> A lot of people out there said we couldn't do this. A lot of people still think we can't. Day by day, our team of incredible scientists, engineers, and operators will continue to prove them wrong. This new EnergyX facility is a critical step towards becoming the largest lithium producer in the world and doing our small part in transitioning the earth to a clean, sustainable energy economy. However, much work still needs to be done, and you can come find me driving innovation right here in Austin.

Kandersteg
December 2024

ACKNOWLEDGMENTS

As is often the case, this book was made possible through a confluence of circumstances, the generosity of numerous people and institutions, and the cherished support of those closest to me. What makes *Extracting the Future* unique, however, in relation to my other research and writing projects, is the fact that the early years of research coincided with first a coup d'état in Bolivia, then the tragedy of the global Covid-19 pandemic, which descended on the country with particular cruelty. Despite the losses, most Bolivians, including many long-term friends and colleagues, managed to survive both historic shocks. But there is no question that these overlapping crises reframed the research project in the context of what is really important: the safety, security, and health and well-being of family and friends.

That said, I am grateful to acknowledge the different ways in which this trajectory over the last five years has brought me to this point. Everything began with a major grant from the Swiss National Science Foundation, which allowed me to put in motion a project that had until then remained only an abstraction. The SNSF continues to invest heavily in fundamental research and the emerging careers of especially junior researchers, something that is becoming increasingly rare internationally. While preparing the grant proposal for the SNSF, I received outstanding research assistance from Jonas Köppel, who went on to conduct his own ethnographic research on lithium in Bolivia as a doctoral student at the Geneva Graduate Institute. I also received supplementary funding between 2018 and 2023 from the Institute of Social Sciences and the Faculty of Social Sciences at the University of Lausanne, for which I am, as always, deeply appreciative.

Research funding allowed me to hire two key project members: David Schröter, who conducted doctoral research in Bolivia, and Zeynep Oguz, who was hired as a senior postdoctoral researcher with expertise in the anthropology of energy, climate policy, and extractivism. We collaborated closely as a team over the years and their work and ideas are reflected in ways both obvious and subtle in the book. I am especially pleased that Zeynep was offered an assistant professorship at the University of Edinburgh in her areas of specialty at the end of her postdoctoral contract. Both David and Zeynep have brilliant careers ahead of them, and I will always wish them the very best.

The research project also benefitted from the wise counsel of a scientific advisory board, which provided an invaluable sounding board for a wide range of questions, from methodology to research ethics. For their time and insights, I express a warm thanks to the following members: Simone Abram, Dominic Boyer, Rebecca Empson, Bret Gustafson, Tobias Haller, Cymene Howe, Marc Hufty, Stefan Leins, and Rita Kesselring.

Between 2019 and 2024, I was fortunate to be invited to discuss my research on lithium, energy transition, and climate politics with many colleagues and institutions. Each event, each Q & A, each critical comment, helped shape my thinking about the unfolding project and its implications. I will always be grateful for the chance to benefit from this collective intelligence and willingness to engage with what was, for the most part, a work-in-progress. My sincerest thanks go out to my hosts, colleagues, and participating students at the following: the University of Sydney (Department of Anthropology, August 2019); Australian National University (National Centre for Indigenous Studies, September 2019); Université Libre de Bruxelles (Laboratoire d'Anthropologie des Mondes, December 2019); the Max Planck Institute for Social Anthropology (February 2020); Laval University and the University of Ottawa (Centre for Indigenous Conservation and Development Alternatives, April 2021); UCL (Department of Anthropology and the UCL Anthropocene Initiative, March 2022); Durham University (Department of Anthropology and the Durham Energy Institute, March 2022); the University of St. Andrews (Department of Social Anthropology and the Centre for Energy Ethics, March 2022); Hranicar Cultural Center (Ústí nad Labem, Czech Republic, April 2022); the Czech Academy of Social Sciences (Sociological Institute, April 2022); the University of New York in Prague (April 2022); Charles University (Faculty of Social Sciences and the Czech Association for Social Anthropology, April 2022); the London School of Economics (Department of Social Anthropology, September 2022); the University of Oslo (Department of Social Anthropology, October 2022); Ludwig Maximillian University–Munich (Institute of Social and Cultural Anthropology, November 2022); the University of St. Gallen (Center for Latin American Studies, March 2023); Université Libre de Bruxelles (Laboratoire d'Anthropologie des Mondes, April 2023); Sciences Po Paris (Center for International Studies and the Interdisciplinary Workshop on Environmental Research, June 2023); the University of Sussex (Department of Anthropology and the Centre for Global Political Economy, October 2023); the University of Cambridge (Department of Social Anthropology, November 2023); Queen's University Belfast (School of History, Anthropology, Philosophy and Politics and the Centre for Sustainability, Equality and Climate Action, December 2023);

the University of Oxford (School of Anthropology and Museum Ethnography, February 2024); the University of Cambridge (Centre of Latin American Studies, February 2024); and Humboldt University (Centre Marc Bloch, November 2024).

The events at the University of Sussex, Queen's University Belfast, and the University of Cambridge (February 2024) were designated as "Leverhulme Lectures," which allows me to acknowledge the second most important circumstance after the financial support from the SNSF that made the underlying research possible. Between September 2023 and August 2024, I had the very good fortune to spend my yearlong sabbatical at the University of Oxford as a Leverhulme Trust Visiting Professor. I express my deepest appreciation to the Leverhulme Trust for this opportunity and also give thanks to its grants officer, Alison Rees, who provided invaluable support both during and after the application process. These positions are awarded through an application by colleagues at the host institution itself, and I will be forever grateful to the small group at Oxford who spearheaded the effort, including David Pratten, Javier Lezaun, and especially Morgan Clarke.

It's simply not possible to individually name all the colleagues at Oxford across multiple schools and faculties whose suggestions and perspectives played a role in shaping my thinking and analysis over the year, during which almost all of this book was written. But I would be remiss to not give explicit thanks for the endless resource that is the Bodleian Libraries, a bibliographic wellspring from which I drank deeply throughout the year. I also became a member of St. Antony's College during my time at Oxford and thank David Pratten (again) and Leigh Payne, in particular, for making my transformation into a lifelong Antonian so congenial.

In Bolivia, as always, my debts run far and wide. A series of officials across the Bolivian government, especially at YLB, were generous enough to grant research access to a number of facilities over the years, including the industrial plants at the salar and the battery center at La Palca. Among the project's many partners in Bolivia, we were fortunate to have Rénan Soruco as the project's main technical advisor. Soruco, who is a professor of chemistry and director of the Lithium Institute at Potosí's Universidad Autónoma Tomás Frías, provided incomparable technical assistance and information at key moments—and in language that could be understood by anthropologists! Among the many Bolivian interlocutors who shared their time and experiences, and who also appear in the book, several, in particular, deserve special mention: Giovana Díaz Ávila, who taught me much about both lithium-ion batteries and the politics of lithium in Bolivia; Carlos Soruco, who opened the doors of Quantum to me and who was always willing to share his vision for what he conceived to be a uniquely Bolivian energy transition; and Juan Carlos Montenegro, whose tragic end does nothing to diminish his life or his career as a pioneering Bolivian lithium scientist and administrator. Finally, I must acknowledge the brilliant transcriptions provided by Catalina Wins Porta, a

Uruguayan anthropologist living in La Paz who brought an expert eye (and ear) to my interviews, including those conducted in English outside of Bolivia.

Speaking of which, I feel a clear sense of obligation to Teague Egan, who generously granted me permission to conduct research with EnergyX at its (now former) headquarters and research facility in Austin, Texas. He was unabashedly enthusiastic about what the company was trying to accomplish as a renegade green energy start-up competing with much larger companies. He also had a sharply defined sense of the importance of the entrepreneur at the center of American—and global—economic life. Although I am critical in this book of the links between the energy transition and existing patterns of production, investment, and consumption, this should not be taken as a critique of EnergyX itself, or of Teague Egan. He didn't have to let an inquisitive anthropologist conduct research from the inside of his company. But he did, and for that I express my deepest thanks.

This book would never have gotten to this point without the early and enthusiastic support of my editor at the University of California Press, Kate Marshall. Kate is a true professional with an editorial touch that manages to be both light and directive at the same time. Her judicious interventions at different stages in the project always led to an improvement, or a necessary course correction, or an important reconsideration. I also benefitted enormously from the editorial work of Chad Attenbourgh, who provided excellent guidance and counsel through the process of production. Artemis Brod brought a masterful eye to bear on the manuscript during copyediting. Finally, I owe a huge debt of gratitude to the three reviewers of the manuscript, whose deep engagement with all aspects of the text will never be forgotten. Although I fear that the final version will still not live up to the high expectations for it imagined by these superlative colleagues, I did my best to treat their recommendations with as much care and attention as I could muster at the end of what was a very long process indeed.

The most important acknowledgement is to my family, whose love, support, tolerance, and understanding are the building blocks of my life. My brother-in-law, Tor Bennström, accompanied me during the last weeks of research in 2023, which included an extended tour of southwest Bolivia beyond the Salar de Uyuni. My wife, Romana, agreed to hold down the fort during my long absences between 2019 and 2023, which included very difficult periods of time during the Covid-19 pandemic. And during the academic year 2023–2024, she returned regularly to Lausanne from Oxford in order to attend to family matters, all of which cut into her own writing time. Without Romana, none of this would have been possible.

NOTES

INTRODUCTION

1. Wright, "Lithium Dreams."
2. IPCC, *Special Report*.
3. European Commission, *A European Green Deal*.
4. Mitchell, *Carbon Democracy*.
5. In 2023, the International Energy Agency (IEA) issued a report that showed that global fossil fuel demand would peak in 2030, after which green energy supplies would come to supplant oil, gas, and coal, giving rise to a "new age of electricity." Twidale, "'Age of Electricity' to Follow Looming Fossil Fuel Peak." Not surprisingly, the IEA report was rejected by the Organization of Petroleum Exporting Countries (OPEC), which called on countries to invest *trillions* of dollars in new oil projects to avoid "energy and economic chaos." World Economic Forum, "IEA."
6. USGS, "Lithium Statistics and Information."
7. Quezada and Carvajal, *Salar de Uyuni*.
8. See, e.g., Risacher, *Estudio económico del Salar de Uyuni*; Risacher and Fritz, "Bromine Geochemistry"; and Risacher and Fritz, "Origin of Salts and Brine Evolution." On the history of ORSTOM, see Bonneuil and Petitjean, "Science and French Colonial Policy." I return to the question of Risacher's work in Bolivia in more detail in chapter 1.
9. USGS, *Mineral Commodity Summaries (Lithium, 2020)*.
10. By 2024, as a function of the global rush to discover new lithium deposits, the USGS had revised its estimates. It now claims that the United States has fourteen million tons of "measured and indicated" lithium reserves across "continental brines, claystone, geothermal brines, hectorite, oilfield brines, and pegmatites." According to these same estimates, Bolivia's reserves had only increased to twenty-three million. USGS, *Mineral Commodity Summaries (Lithium, 2024)*. Nevertheless, by including the ten million likely additional tons, Bolivia alone would still hold almost 30 percent of the world's lithium reserves.

11. Here, let me acknowledge the many ways in which the concept of the Anthropocene has been contested from a variety of angles. In her characteristically playful review of some of these debates, Donna Haraway nevertheless recognizes that "immense irreversible destruction is really in train, not only for the 11 billion or so people who will be on earth near the end of the 21ˢᵗ century, but for myriads of other critters too." Haraway, "Anthropocene, Capitalocene, Plantationocene, Chthulucene," 161. I use *Anthropocene* throughout the book to signal both the collective problems of interdependent life under these conditions and the "stories (and theories)" that are being told about these conditions. Thus, I agree with Haraway that "Anthropocence" is "just big enough to gather up the complexities and keep the edges open . . . for surprising new and old connections." "Anthropocence," 161.

12. See, e.g., Babidge and Bolados, "Neoextractivism and Indigenous Water Ritual in Salar de Atacama, Chile"; Bustos-Gallardo, Bridge, and Prieto, "Harvesting Lithium"; Jerez Henríquez, Garcés, and Torres, "Lithium Extractivism and Water Injustices in the Salar de Atacama, Chile"; Romero, Méndez, and Smith, "Mining Development and Environmental Injustice in the Atacama Desert of Northern Chile"; and Blair, Balcázar, Barandiarán, and Maxwell, "The 'Afterlives' of Green Extractivism."

13. See, e.g., Abelvik-Lawson, "Indigenous Environmental Rights, Participation and Lithium Mining in Argentina and Bolivia"; Fornillo, "La energía del litio en Argentina y Bolivia"; Göbel, "La minería del litio en la Puna de Atacama"; López, Obaya, Pascuini, and Ramos, *Litio en la Argentina*; Obaya, López, and Pascuini, "Curb Your Enthusiasm"; and Schiaffini, "Litio, llamas y sal en la Puna Argentina."

14. See, e.g., Argento, "Espejo de sal"; Arze, "Abandono del discurso estatista en la industrialización del litio,"; Calla et al., *Un presente sin futuro*; Daza, "Historia del extractivismo del litio en Bolivia"; Montenegro, "El modelo de industrialización del litio en Bolivia"; Nacif, "Bolivia y el plan de industrialización del litio"; Obaya, *Estudio de caso sobre la gobernanza del litio en el Estado Plurinacional de Bolivia*; Olivera, *La industrialización del litio en Bolivia*; Poveda, *Industrialización del litio en Bolivia*; Revette, "This Time It's Different"; Rodríguez-Carmona and Aranda, *De la salmuera a la batería*; Sanchez-Lopez, "From a White Desert to the Largest World Deposit of Lithium"; Sanchez-Lopez, "Territory and Lithium Extraction"; Solón, *Espejismos de abundancia*; and Szoke, "The Lithium Economy."

15. See, e.g., Ahmad, "The Lithium Triangle"; Barandiarán, "Lithium and Development Imaginaries in Chile, Argentina and Bolivia"; Heredia, Martinez, and Surraco Urtubey, "The Importance of Lithium for Achieving a Low-Carbon Future"; Kingsbury, "Lithium's buzz"; Seefeldt, "Lessons from the Lithium Triangle"; and Soto Hernandez and Newell, "Oro blanco."

16. Because the academic literature on extractivism across disciplines is enormous, it would make little sense to attempt to provide a dissertation-ish compendium here. Instead, I will refer to specific works—here and throughout the book, on extractivism or otherwise—when they are particularly relevant, except when a brief survey of sources is justified to orient the reader or provide background beyond what can be accomplished in the main text.

17. Jobson, Gómez-Barris, Howe, and Winchell, "Extractivism's Limits."

18. Unfortunately, this book was too far into production for me to engage with the arguments and perspectives in Thea Riofrancos's much-anticipated *Extraction: The Frontiers of Green Capitalism*, which will undoubtedly become an indispensable addition to the literature.

19. Anthias, "Contested Energy Futurities at Bolivia's Hydrocarbon-Conservation Frontier." For an analysis of how deeply embedded political economic structures can stifle the emergence of "post-extractivist" movements, see Andreucci and Radhuber, "Limits to 'Counter-Neoliberal' Reform."

20. Djukanović, "How Does an Extractivist Frontier Come to Being?"

21. See, e.g., Gudynas, "Diez tesis urgentes sobre el nuevo extractivismo."

22. See, e.g., Svampa, *Neo-Extractivism in Latin America*.

23. Voskoboynik and Andreucci, "Greening Extractivism."

24. Dunlap and Riquito, "Social warfare for lithium extraction?"

25. Howe and Boyer, "Aeolian Extractivism and Community Wind in Southern Mexico." See also Jessica Smith's study of corporate social responsibility programs among US mining, oil, and gas companies, in which engineers are forced to reconcile competing interests—from the public, from their companies, and from their professions—in the effort to "extract accountability." Smith, *Extracting Accountability*. For a study that argues that solar power projects have an essential role to play in the promotion of environmental justice, despite the problems of green extractivism, see Mulvaney, *Solar Power*.

26. Harvey, *The Condition of Postmodernity*.

27. For an important study of how conflicts around competing visions of resource extraction play out beyond the gaze of state policymaking, see Riofrancos, *Resource Radicals*.

28. Goodale, *A Revolution in Fragments*.

29. See Winchell, "Fire's Alter-Lives."

30. Despite the impact of both a coup d'état in Bolivia (see chapter 2) and the global Covid-19 pandemic, I managed to complete eight and a half months of ethnographic fieldwork in Germany, Bolivia, and the United States between July 2019 and August 2023. Primary research consisted of observations at a broad array of institutions, sites of extraction, manufacturing facilities, laboratories, and research and development offices, among other locations; recorded and unrecorded interviews with over a hundred people; and documentary analysis and archival research.

31. See Goodale, "Futures Fixed and Foreclosed," "Lithium Scale-Making and Extractivist Counter-Futurities in Bolivia," and "Moving the Bones."

32. Besides myself as director, the other core members of the research team were David Schröter, who conducted doctoral research throughout the salar region but primarily in the village of Río Grande (see Schröter, "The Future is Now"), and Zeynep Oguz, the project's senior postdoctoral researcher (now an assistant professor of anthropology at the University of Edinburgh), who oversaw data analysis and the project's insertion into wider disciplinary and interdisciplinary networks.

33. See Nader, "Studying Up." See also Gusterson, "Studying Up Revisited."

34. Marcus, "Ethnography in/of the World System."

35. Merry, *Human Rights and Gender Violence.*

36. *Human Rights and Gender Violence*, 29.

37. Goodale, "Futures Fixed and Foreclosed."

38. Hecht, "Interscalar Vehicles for an African Anthropocene," 135.

39. "Interscalar Vehicles for an African Anthropocene," 114. See also Xiang, "Multi-Scalar Ethnography."

40. Yarrow, "Remains of the Future."

41. Nixon, *Slow Violence and the Environmentalism of the Poor.*

42. For examples of the important work that Indigenous scholars themselves are doing to reorient debates around the Anthropocene and climate change mitigation, see Davis and Todd, "On the Importance of a Date," and Whyte, "Indigenous Climate Change Studies."

43. Tsing, *The Mushroom at the End of the World*, 28.

CHAPTER 1

1. Gustafson, *Bolivia in the Age of Gas.*

2. Yarrow, "Remains of the Future"; Gordillo, *Rubble.*

3. Palsson and Swanson, "Down to Earth."

4. Goodale, "Of Crystals and Semiotic Slippage."

5. Palm oil: Li and Semedi, *Plantation Life*; water: Ballestero, *A Future History of Water*; broccoli: Fischer and Benson, *Broccoli and Desire*; tobacco: Benson, *Tobacco Capitalism*; the matsutake mushroom: Tsing, *The Mushroom at the End of the World*; sugar: Mintz, *Sweetness and Power.*

6. The Bodleian Libraries of the University of Oxford, which comprise twenty-six separate libraries and thirteen million printed volumes.

7. Bednarski, *Lithium*; Kunasz, *The Lithium Legacy.*

8. Brown, *Lithium*; see also, e.g., De Moore and Westmore, *Finding Sanity.*

9. Weeks and Larson, "J.A. Arfwedson and His Services to Chemistry." This section draws on the short biography of Arfwedson cowritten by Mary Elvira Weeks, who was herself an intriguing historical figure. Weeks (1892–1975) was the first woman to earn a PhD in chemistry from the University of Kansas and also the first woman to secure a faculty position in chemistry at the university. During her time at Kansas, where she eventually was promoted to associate professor, she published her magisterial *Discovery of the Elements*, in which she wrote short histories of different elements, a work that became a standard text in chemistry through seven editions between 1934 and 1968. See Bray, "KU's First Woman of Chemistry, Mary Elvira Weeks."

10. Quoted in Weeks and Larson, "J. A. Arfwedson and His Services to Chemistry," 405.

11. Quoted "J. A. Arfwedson and His Services to Chemistry," 407.

12. Shorter, "The History of Lithium Therapy."

13. Brown, *Lithium*.

14. Quoted in Shorter, "The History of Lithium Therapy," 5.

15. Brown, *Lithium*, 91.

16. *Lithium*, 96.

17. Donovan, *Fizz*.

18. Brown, *Lithium*.

19. *Lithium*, x.

20. *Lithium*.

21. *Lithium*, xi.

22. Lowe, "I Don't Believe in God, but I Believe in Lithium."

23. Parsons and Zaballa, *Bombing the Marshall Islands*, 55.

24. Bednarski, *Lithium*.

25. Murray, *Long Hard Road*, xi.

26. *Long Hard Road*, xii-xiii.

27. *Long Hard Road*, 16.

28. Mitchell, *Carbon Democracy*.

29. Murray, "Who Really Invented the Rechargeable Lithium-Ion Battery?"

30. "Who Really Invented the Rechargeable Lithium-Ion Battery?"

31. "Who Really Invented the Rechargeable Lithium-Ion Battery?"

32. Bednarski, *Lithium*, 45.

33. Galeano, *Open Veins of Latin America*.

34. *Open Veins of Latin America*, 20.

35. Cole, *The Potosí Mita, 1573–1700*.

36. Galeano, *Open Veins of Latin America*, 21.

37. *Open Veins of Latin America*, 22.

38. *Open Veins of Latin America*, 32.

39. *Open Veins of Latin America*, 32.

40. Nash, *We Eat the Mines and the Mines Eat Us*.

41. Mesa, Gisbert, and Mesa Gisbert, *Historia de Bolivia*.

42. *Historia de Bolivia*, 363.

43. Cushman, *Guano and the Opening of the Pacific World*.

44. *Guano and the Opening of the Pacific World*, 64.

45. *Guano and the Opening of the Pacific World*, 68.

46. *Guano and the Opening of the Pacific World*, 73. In the years before the First World War, the German chemists Fritz Haber and Carl Bosch figured out how to produce nitrogen from the atmosphere. The invention of the Haber-Bosch process made the synthetic production of nitrates possible, a shift that eventually rendered the global market in guano and saltpeter obsolete. The German nitrate industry played a major role in the production of various kinds of explosives used by Germany during the war.

47. See Goodale, *A Revolution in Fragments*.

48. Mesa, Gisbert, and Mesa Gisbert, *Historia de Bolivia*, 614. Although I focus here on particular conflicts over the terms of extractivist entanglement, those that clearly bear on the ethnographic study of Bolivia's lithium project, there are other

relevant episodes in the wider resource chronology that I have had to leave out largely for reasons of space. For example, continuing the association between resource extraction and territorial loss, Bolivia fought an intermittent war with Brazil between 1899 and 1903 for control over a disputed region of the Amazon that had abundant rubber trees (used to produce tires for the emerging global automobile market). After being defeated by Brazilian forces, Bolivia was forced to sign the Treaty of Petrópolis, which ceded a large section of northwestern Bolivia to Brazil. And between 1932 and 1935, Bolivia clashed with Paraguay in the disastrous Chaco War, during which Bolivia lost around sixty-five thousand soldiers and over fifty thousand square kilometers of territory in the oil- and gas-rich Chaco region. Although the causes of the Chaco War are complex, it was widely believed that it was a proxy war between Standard Oil of New Jersey, which backed Bolivia, and Royal Dutch Shell, which backed Paraguay, a belief that in part led the postwar Bolivian government to create *Yacimientos Petrolíferos Fiscales Bolivianos* (YPFB), Bolivia's state oil and gas company. Klein, *Bolivia*, 185.

49. Whyte, *The Morals of the Market.*

50. *Historia de Bolivia*, 623. It should be noted that one of these Bolivian historians, Carlos Mesa Gisbert, was actually the vice president during the second Goni administration. Although the Gas War led to the resignation and exile of Sánchez de Lozada, Mesa had managed to distance himself from the actions of the government just early enough to avoid being cast into the abyss himself—at least for the time being. In 2005, while serving as caretaker president, Mesa was forced to resign in the months before the historic elections of 2005, which brought Evo Morales and the MAS to power.

51. Lazar, *El Alto, Rebel City.*

52. US Department of the Interior, "Earth's Resources to be Studied From Space." For the definitive study of how the US Department of the Interior has long functioned as an instrument of American imperial power beyond the country's borders, particularly in relation to resources considered critical to US economic and security interests, see Black, *The Global Interior.*

53. US Department of the Interior, "Earth's Resources to be Studied From Space."

54. See Carter, "ERTS-A" and Carter and Rowan, "Applying Satellite Technology to Energy and Mineral Exploration."

55. Stoertz and Carter, "Hydrogeology of Closed Basins and Deserts of South America."

56. Carter, "ERTS-A."

57. Ericksen, Vine, and Ballón, *Lithium-Rich Brines at Salar de Uyuni and Nearby Salars in Southwestern Bolivia.*

58. Davis et al., *Progress Report on Lithium-Related Geologic Investigations in Bolivia.*

59. Ericksen, Vine, and Ballón, *Lithium-Rich Brines at Salar de Uyuni and Nearby Salars in Southwestern Bolivia.*

60. *Lithium-Rich Brines at Salar de Uyuni and Nearby Salars in Southwestern Bolivia*, 3.

61. Kunasz, *The Lithium Legacy*.

62. Zuboff, *The Age of Surveillance Capitalism*.

63. NASA, "Landsat Science."

64. Nacif, "Bolivia y el plan de industrialización del litio."

65. Bonneuil and Petitjean, "Science and French Colonial Policy."

66. Quezada and Carvajal, *Salar de Uyuni*. See also the introduction.

67. They collaborated in a 1978 expedition; see Davis et al., *Progress Report on Lithium-Related Geologic Investigations in Bolivia*.

68. As far as I know, the first time the concept of the fiscal reserve was used by the Bolivian government was in the Law of December 5, 1917, a legal starting point that is invoked repeatedly in the jurisprudence of Bolivia's lithium project. More broadly, Bolivian governments throughout the twentieth and into the twenty-first century made use of the fiscal reserve dozens of times to assert control over oil, gas, and mining resources, in addition to lithium.

69. The law that created CIRESU (Law 719) also made a concession to other extractive activities on the Salar de Uyuni, most importantly artisanal salt mining, which was traditionally centered around the community of Colchani. See Schröter, "The Future is Now." Although Colchani continues to mine—or, better, harvest (see Bustos-Gallardo, Bridge, and Prieto, "Harvesting Lithium")—salt, its economy depends much more now on the salar tourism industry.

70. This section draws on the detailed history of this period compiled by the sociologist Federico Nacif. Nacif, "Bolivia y el plan de industrialización del litio."

71. Orellana Rocha, "El litio."

72. Abelvik-Lawson, "Indigenous Environmental Rights, Participation and Lithium Mining in Argentina and Bolivia," 228.

73. Orellana Rocha, "El litio."

74. Bednarski, *Lithium*, 130–131.

75. Yarrow, "Remains of the Future," 568.

76. "Remains of the Future," 568.

77. Ballestero, *A Future History of Water*.

78. Gordillo, *Rubble*.

79. Robbins, "The Present and the Future in the Present."

80. Anthias, "Contested Energy Futurities at Bolivia's Hydrocarbon-Conservation Frontier."

81. See Goodale, "An Anthropology of Impossible Futures."

82. Sassen, *Explusions*.

CHAPTER 2

1. Niezen, *The Origins of Indigenism*; Honneth, *The Struggle for Recognition*; Fraser, "From Redistribution to Recognition?"

2. Honneth, *The Struggle for Recognition*.

3. Goodale, "Reclaiming Modernity."

4. Tobar, "Revolt on High."

5. Gustafson, *Bolivia in the Age of Oil*.

6. For an innovative recent analysis of the rise and ultimate fall of the Morales government, see McNelly, *Now We Are in Power*, which argues that this historic period should be understood as a "passive revolution" marked by inherent contradiction and prolonged crisis.

7. Prada, "Bolivian Nationalizes the Oil and Gas Sector."

8. For more on the separatist movement and antigovernment mobilizations during these years, see Calla, "Making Sense of May 24ᵗʰ in Sucre"; Fabricant and Postero, "Contested Bodies, Contested States"; Goodale, *A Revolution in Fragments* and "God, Fatherland, Home"; and Gustafson, "Spectacles of Autonomy and Crisis." A broader historical analysis of the long-standing "autonomist discourse" in Bolivia can be found in Plata, "El discurso autonomista de las élites de Santa Cruz."

9. Lallemand, "Il est soupçonné de trafic d'acide à destination des narcos."

10. Nacif, "Bolivia y el plan de industrialización del litio." See also Schröter, "The Future is Now."

11. Bednarski, *Lithium*, 96.

12. Nacif, "Bolivia y el plan de industrialización del litio."

13. Quoted in Nacif, "Bolivia y el plan de industrialización del litio."

14. Goodale, *A Revolution in Fragments*.

15. Camba ethnonationalism is a movement centered in Santa Cruz that has historically advanced the argument that lowland Bolivians share a single ethnicity that is distinct from that of the highlands, which is represented as a single "Colla" ethnicity. For a typical presentation of the so-called two Bolivias hypothesis, see Antelo Gutiérrez, *Los Cambas*; for an anthropological critique of the "performative politics" of Camba ethnonationalism, see Fabricant, "Performative Politics."

16. Although lithium and other "evaporitic resources" are usually described as "non-metallic" in government documents during these years, it must be remembered (as was seen in chapter 1) that lithium is, in fact, an alkali metal. This perhaps unintentional semiotic slippage (see Goodale, "Of Crystals and Semiotic Slippage") underscores just how "indeterminate" (see Weszkalnys, "Geology, Potentiality, Speculation") lithium was within the imaginaries of MAS policymakers and legislators during most of the 2000s and 2010s—that is, until more precise knowledge about lithium itself became a governmental priority (and technological imperative).

17. Nacif, "Bolivia y el plan de industrialización del litio."

18. PA25 was printed as a pamphlet by the government and distributed widely under the title "Agenda Patriótica 2025: Trece Pilares de la Bolivia Digna y Soberana."

19. See Goodale, "Timerendering." See also Gutiérrez Aguilar, *Rhythms of the Pachakuti*.

20. McNeish, *Sovereign Forces*.

21. GNRE, *Memoria 2015*.

22. Quispe, "Hasta 2019, Bolivia invertirá $us 925 MM en la industria del litio."

23. See Goodale, *A Revolution in Fragments*.

24. Farthing and Becker, *Coup*, 62.

25. Vice Ministry of Communication, "Presidente Morales inaugura la planta industrial de cloruro de potasio en Uyuni."

26. Bednarski, *Lithium*.

27. Not only does Germany have by a considerable margin the largest economy by nominal GDP in the EU, but three of the ten largest private companies by revenue across the entire EU are German automakers (Volkswagen, Mercedes-Benz, and BMW).

28. Bednarski, *Lithium*, 139.

29. *Lithium*, 139.

30. *Lithium*, 139. See also Mitra and Nienaber, "In the New Lithium 'Great Game,' Germany Edges Out China in Bolivia."

31. In a twist of geographical fate, the small town of Zimmern ob Rottweil is located in the far southwestern corner of Germany in a way that very closely approximates the relative location of the Salar de Uyuni in Bolivia.

32. Mitra and Nienaber, "In the New Lithium 'Great Game,' Germany Edges Out China in Bolivia" (emphasis added).

33. It is important to note that YLB signed an agreement with another Chinese company in early 2019 that adopted several of the same provisions as the YLB-ACISA contract. However, the agreement with Xinjiang TBEA Group was also different in key respects. The contract gave the Chinese company the right to explore the feasibility of lithium production in two smaller salares, Coipasa and Pastos Grandes, but the project was framed very much for the long-term and lacked many of the specificities of the YLB-ACISA agreement. It also received little attention at the time and, unlike the YLB-ACISA agreement, did not become a source of conflict in Potosí.

34. García Linera, "Un Potosí federal dejará de recibir Bs 3.900 millones del Estado."

35. Jemio, "Una Bolivia en transición política repiensa cómo industrializar su litio."

36. Farthing and Becker, *Coup*.

37. *Coup*.

38. Grisaffi, "A Brief History of Coca."

39. As a Fulbright scholar (to Romania) in 2003–2004, I came to learn much about the inner workings of the US diplomatic system, since Fulbright scholars are supervised by the US State Department. During the period of the scholarship, we were given badges that allowed us to enter the US Embassy in Bucharest, shop at the commissary, use the diplomatic pouch for mail, and otherwise get to know the "FSOs," or foreign service officers, which is how people working for US embassies refer to themselves.

40. As this conversation obviously wasn't recorded, or even planned, I had to reconstruct it from memory based on notes I took after arriving in La Paz the next day.

41. This unexpected encounter ended with the defense attaché handing me his business card. I had said in passing that I was uneasy about traveling around the country in light of the violence and political upheaval. He said I could call him directly if I needed any help. I was both surprised to be reassured by this offer but also somewhat aghast about returning to Bolivia during a coup with a new connection—however fleeting—with an American intelligence officer.

42. ANF, "Pobladores del sudoeste potosino declaran emergencia y rechazan a nuevo gerente de la estatal de litio."

43. Not surprisingly, even the question of evaporation rates takes on political dimensions, and not only in Bolivia. YLB technicians would typically vastly underestimate the timeframes; for example, I was told during an early period in my research that salt compounds could be derived in "four to five months." However, proponents of alternative extraction technologies tend to overestimate evaporation timeframes in order to highlight the benefits of their proposals (see chapter 6).

44. Weszkalnys, "Geology, Potentiality, Speculation."

CHAPTER 3

1. Molina Rivero, *De memorias e identidades*

2. Tellería, "La hegemonía verde."

3. It should be remembered that this was at the height of the lucrative guano and saltpeter trade in Bolivia, in which the British government and London financiers were heavily invested. See chapter 1.

4. Bowman, "Results of an Expedition to the Central Andes."

5. "Results of an Expedition to the Central Andes," 178. In fact, as we have seen, Minchin was alive in 1914 and living in London, although he had suffered a stroke by the time of Bowman's article.

6. "Results of an Expedition to the Central Andes," 180. Bowman spells Ballivián's name without using the diacritic.

7. Bowman went on to have his own complex professional career in the decades after his early research in Bolivia. He was a key advisor to both Woodrow Wilson and Franklin D. Roosevelt during the First and Second World Wars. But as the geographer Neil Smith explains in his biography of Bowman, "Roosevelt's geographer" was also a notorious antisemite who hampered efforts to resettle Jewish refugees during the Holocaust. And as president of Johns Hopkins University, he implemented formal anti-Jewish quotas in admissions after the war and took other steps to marginalize Jewish faculty and students at the major US research university. Smith, *American Empire*.

8. Servant and Fontes, "Les lacs quaternaires des hauts plateaux des Andes boliviennes." Two of the smaller lakes that were created from the former Lake Minchin were Lake Poopó and Lake Uru Uru, both located in what is today Oruro Department to the north of the Salar de Uyuni. For much of my research career in Bolivia, Lake Poopó still had water, which flowed from Lake Titicaca farther to the

north via the Desaguadero River. But by about 2016, the lake had evaporated, in large part due to changing weather patterns caused by climate change. Lake Uru Uru, which is on the outskirts of the city of Oruro, still retains some water, but it is also likely to evaporate completely over the next decade. Most of the diminishing lake is now a large toxic waste dump filled with plastic and other refuse from the city as well as iron oxide and hydroxide runoff from several local mines.

9. By comparison, several of the other largest salares in the world include Salinas Grandes in Argentina (6,000 sq km), the Etosha Pan in Namibia (4,800 sq km), and the Salar de Atacama (now in Chile; 3,000 sq km).

10. Borsa, et al., "Topography of the Salar de Uyuni, Bolivia from Kinematic GPS."

11. See Hand, "The Salt Flat with Curious Curves."

12. Davis et al., *Progress Report on Lithium-Related Geologic Investigations in Bolivia*, 7.

13. Quezada and Carvajal, *Salar de Uyuni*, who draw on Houston et al., "The Evaluation of Brine Prospects and the Requirement for Modifications to Filing Standards."

14. Quezada and Carvajal, *Salar de Uyuni*, 5.

15. It is also worth noting that the Salar de Atacama's brine is more lithium-rich than the Salar de Uyuni's: 2.55 grams per liter on average vs. 0.42.

16. In his *Requiem for a Republic*, which was also his own requiem of sorts, given that he died tragically at the age of thirty-nine, a year before it was published, the Bolivian political essayist Sergio Almaraz Paz offers an indelible portrait of the suffering and toxic working conditions of Bolivian miners in a chapter entitled, appropriately, "Mines as Cemeteries." Almaraz Paz, *Réquiem para una república*.

17. In an earlier publication, I write somewhat more expansively about this astonishing exchange with the YLB geologist. Goodale, "Of Crystals and Semiotic Slippage."

18. Marston, "Strata of the State."

19. Marston, *Subterranean Matters*.

20. Nash, *We Eat the Mines and the Mines Eat Us*.

21. Critical resource geographers have examined in detail the concept of materiality in relation to the contested concept of natural resources and economic transformation. For an essential review of these debates, see Bakker and Bridge, "Material Worlds Redux." See also Bridge, "Material Worlds."

22. See Ingold, *Being Alive*.

23. Quezada and Carvajal, *Salar de Uyuni*, 14. By comparison, the world's deepest borehole remains the Kola Superdeep Borehole, a Soviet project undertaken during the Cold War above the Arctic Circle. The borehole reached a depth of almost 12,300 meters.

24. As with my research among the salar's ojos with the head geologist, I also examine different aspects of this demonstration of core samples, and my discussion with the assistant geologist, in Goodale, "Of Crystals and Semiotic Slippage."

25. See Farthing and Becker, *Coup*.

26. In the October 2020 elections, the MAS ticket received about 55% of the vote, considerably more than Morales received in the contested 2019 elections and even more than Morales and García Linera had received in the historic elections of 2005.

27. I examine Bolivia's turn toward DLE in more detail in chapter 6.

28. At the same time that the lithium project was brought under the fold of the new ministry dominated by officials and institutional logics from Bolivia's hydrocarbon industries, the Arce government appointed Franklin Molina Ortiz as its first minister. An economist from Santa Cruz, most of Molina's experience in government service had been in planning and management in Bolivia's hydrocarbon, electricity, and telecommunications sectors.

29. For more on the Bolivian state initiative through which Calderón was able to travel abroad for her master's degree, see the discussion of the 100 Grants program in chapter 4.

30. See Schröter, "The Future is Now."

31. See Pauline Destrée's study of the emergence of a version of flexible extractivism in Ghana, where energy policymakers have pushed back against demands that they transition away from oil and gas development as a counter-hegemonic assertion of both climate and carbon justice. Destrée, "Gaseous Politics."

32. Nixon, *Slow Violence and the Environmentalism of the Poor.*

33. See Ong, *Flexible Citizenship.*

34. Anthias, "The Pluri-Extractivist State."

35. See also Goodale, "Lithium Scale-Making and Extractivist Counter-Futurities in Bolivia."

36. Li, "Indigeneity, Capitalism, and the Management of Dispossession," 400.

CHAPTER 4

1. And to add to the picture of China's unique position within the various extractive economies that are at the center of the global energy transition, it should be noted that Chinese companies control most of the cobalt production in the Democratic Republic of Congo and, increasingly, in Indonesia, where cobalt mining is growing rapidly along with the existing nickel industry, also dominated by Chinese companies. Although not as indispensable as lithium itself, both cobalt and nickel are key ingredients in some of the most widely used lithium-ion batteries for EVs.

2. This assumes that a massive supply of new lithium will be needed to produce batteries and that other sources will not be available. Although lithium scientists are currently working on different ways to recycle lithium from existing batteries, the technical challenges are considerable.

3. Hayden, *The Spectacular Generic.*

4. Quoted in Galeano, *Open Veins of Latin America,* 32.

5. *Los Tiempos,* "Pumari convoca a detractores que quieran insultarle."

6. *Los Tiempos*, "Pumari pide disculpas entre gritos, insultos y tomatazos."

7. For what it's worth, Bolivia doesn't have significant reserves of cobalt, or for that matter, of nickel, another "critical" mineral widely used in the production of lithium-ion batteries.

8. Female scientists and researchers have been at the center of Bolivia's lithium project from the beginning. Although lithium research was not part of this 2006 congress, it is worth noting that the two students to receive honorable mention in the materials science category were, like Díaz, also young women. Soledad Yanarico Gutiérrez (UMSA) was recognized for her paper "Comparison of Methodologies for Determining the Corrosion Rate of Low Carbon Steel," while Susan Gonzáles Soto from Cochabamba's *Universidad Mayor de San Simón* (UMSS) was awarded honorable mention for her study of the "Separation of Sulfur by Flotation from Volcanic Minerals of the Western Cordillera." Velasco Hurtado, "VII congreso nacional de metalurgia y ciencia de materiales."

9. The other was Cesario Ajpi Condori, a graduate in chemistry from UMSA. Ajpi later completed a PhD in chemical engineering from the KTH Royal Institute of Technology in Sweden based on a study of "hybrid materials for lithium-ion batteries."

10. A version of this technology would later be part of a new approach to the production of lithium carbonate from brine called "direct lithium extraction," or DLE. See chapter 6.

11. Ayub and Hashimoto, *The Economics of Tin Mining in Bolivia*, 77.

12. *The Economics of Tin Mining in Bolivia*, 77.

13. *The Economics of Tin Mining in Bolivia*, 77.

14. *The Economics of Tin Mining in Bolivia*, 77.

15. For more on the 100 Grants program, see Jonas Köppel's 2023 doctoral dissertation. Köppel, *Lithium Trajectories in Bolivia and beyond*. Köppel was among a small group of mostly anthropologists who conducted research on different dimensions of Bolivia's lithium project beginning around 2019. His project focused on the scientists involved in the project, including many who worked at La Palca.

16. Bednarski, *Lithium*, 138.

17. As Giovana Díaz told me, this explains in large part why Greentech was selected for the project, despite the fact that it was mostly a "PV equipment manufacturer." Its proposal to GNRE/COMIBOL included the training period at the CEA, one of the largest energy research centers in the world, and the extended period of follow-on *capacitación* at La Palca.

18. NobelPrize.org, "The Nobel Prize in Chemistry."

19. Despite an interest in new materials, almost all anodes continue to be based on a carbon material, usually graphite. For the electrolyte, most of the research on alternatives to a liquid medium focuses on different kinds of solid-state batteries, but structural limitations have kept this technology from being widely adopted.

20. The Ten Commandments of Public Service were announced by the Morales government about a year before the fateful 2019 elections. This "decálogo" was meant to be posted in visible locations in all public companies and agencies as a

reminder that the state's employees were part of a bigger political project that went beyond salaries and working conditions. The commandments, which were stated as values, were: compromise, legality, loyalty, probity, truth, efficiency, honesty, impartiality, solidarity, and transparency.

CHAPTER 5

1. Larson, *Cochabamba, 1550–1900*, 3. For a comparative study of Cochabamba as one among several critical production "enclaves" of the Inca Empire, see, La Lone and La Lone, "The Inka State in the Southern Highlands." Laura Gotkowitz's research examines social transformation and political mobilization in the Cochabamba Valley during the first half of the twentieth century leading up to the 1952 National Revolution. See Gotkowitz, *A Revolution for Our Rights*. A fascinating and provocative anthropological reflection on recent struggles over land and history in a more distant part of Cochabamba Department can be found in Mareike Winchell's 2022 book. Winchell, *After Servitude*.

2. For a classic study of Patiño and his family, see Geddes, *Patiño*.

3. Bednarski, *Lithium*, 29.

4. The Bolivian community of northern Virginia is large and vibrant enough to support a kind of "little Bolivia" in the suburbs of Washington, DC, which includes a number of Bolivian restaurants, a Bolivian soccer/football league, and dancing clubs that practice for a local Carnaval parade that resembles the one in Oruro. During the eleven years I lived in Fairfax County while teaching at George Mason University, I frequently came across groups of Bolivian-Americans in high school parking lots who were practicing the familiar dances of the Diabladas, Caporales, Morenadas, and others.

5. The geographer Andrea Marston's body of research provides a comprehensive picture of cooperative mining in Bolivia. See, e.g., Marston, "Alloyed waterscapes," "Subsoil Politics," and especially *Subterranean Matters*.

6. Gotkowitz, *A Revolution for Our Rights*, 99, 318. This massacre took place in the region around Lake Titicaca, the heartland of the Aymara people, during the government of the *caudillo* president Bautista Saavedra. Regional Aymara leaders led a *sublevación*, or uprising, against the hated local elite, including the town officials in Jesús de Machaca. After an Aymara militia killed a number of these officials, including the town *corregidor* and his family, Saavedra ordered a force of about 1,500 Bolivian soldiers to move into the region. They engaged in a scorched-earth campaign against the Aymara communities that involved the killing of hundreds of men, women, and children; the burning of houses; and the destruction of over a thousand livestock. Mesa, Gisbert, and Mesa Gisbert, *Historia de Bolivia*, 444–445. See also Choque Canqui and Ticona Alejo, *Jesús de Machaqa*.

7. One curious complication in the process of launching Quantum involved the assigning of a World Manufacturers Identifier (WMI) to Quantum's first EVs. The WMI is a series of unique numbers that are part of the longer Vehicle Identification

Number (VIN). The Society of Automotive Engineers (SAE), which is based in the US state of Pennsylvania, assigns WMIs to auto manufacturers around the world in an effort to standardize the process of vehicle identification. As Soruco explained somewhat bemusedly, because Quantum was the first Bolivian car company, EV or otherwise, the SAE had never had to issue a WMI for Bolivia. So the SAE assigned Quantum the number used for Venezuelan auto manufacturers as a provisional measure, before agreeing to create a new WMI for Bolivia's first car company.

8. Dube, "Look Out, Tesla, There's a Really Tiny Competitor in Your Rearview Mirror."

9. Just for comparison, the Tesla Model Y, the most popular of the company's EVs, has a 220-kW engine, meaning *220,000* watts.

10. The names of the Quantum owners in this section are pseudonyms.

11. La Angostura is an artificial lake built in 1945 with financing from the government of Mexico. The lake, which was originally used for irrigation in the region, is about twenty kilometers from Cochabamba and a popular destination for weekend excursions.

12. Daggett, "Renewable Masculinities."

13. See Olivera, *¡Cochabamba!*

CHAPTER 6

1. I say salares, plural, because the convocatoria, and the broader plan for DLE in Bolivia, involves not only the Salar de Uyuni, but two other salares, Coipasa and Pastos Grandes.

2. Scholars have already begun to estimate the environmental impacts of the shift from brine evaporation to DLE, with particular concern for what one study describes as the "hydrosocial impacts" of this new technology. See Blair et al., "Lithium and Water." See also Vera, et al., "Environmental impact of direct lithium extraction from brines."

3. For a fascinating study of the epistemic status of the "problem" within engineering, see Abram, "Problem Solving as Selective Blindness."

4. High, "Resisting Energy Transitions."

5. Wright, "The Astonishing Transformation of Austin." By a remarkable coincidence, Lawrence Wright wrote one of the most important early essays on Bolivia's "lithium dreams," also in the *New Yorker* (see introduction). To recall, his essay poses the question, "Can Bolivia Become the Saudi Arabia of the Electric-Car Era?" His answer, in 2010, was probably not.

6. Quoted in Wright, "The Astonishing Transformation of Austin."

7. "The Astonishing Transformation of Austin."

8. "The Astonishing Transformation of Austin."

9. I mention this because I always owned cars of different kinds, including gas-guzzling minivans, during the many years I lived in the United States. When I moved to Lausanne, Switzerland, in 2014, we decided to try and live without cars. Under

circumstances that differ greatly with respect to transportation networks, city geography, and government incentives to buy annual bus and train passes, we are able to do everything we need (or want) to do through public transportation, cycling, walking, intercity lake ferry, and the occasional use of a shared car service called Mobility.

10. Although other countries allow the use of vanity license plates—personalized plates owners pay a fee for—it is particularly associated with drivers across the US. For example, when I was growing up in Southern California, my father had a vanity plate for a number of years that said, simply, "GOODALE."

11. Although we didn't discuss this rap artist by name, it didn't take much detective work to discover that his name is Sammy Adams, aka Boston's Boy (which was also the name of the album released by 1st Round Records in 2010). However, Adams, born Samuel Adams Wisner in Cambridge, Massachusetts, turns out to be a rap star with an unusual background. The white son of an upper-middle-class family, he was a soccer player at the private Trinity College in Connecticut (the alma mater of Tucker Carlson), where he was captain of the varsity team during his senior year, two years before he met Teague on the beach at Nantucket.

12. Balch, "Punta del Este."

13. As I was finalizing the manuscript, I learned that von Vacano was working on his own book about Bolivia's lithium project. Written from his insider's perspective as an informal advisor to Luis Arce, its working title is "Power over Energy: Lithium, Bolivia, and the Global Struggle for the New Green Economy."

14. For another study of the ways in which energy entrepreneurs are often driven by moral ambitions to view their projects as forms of quasi-religious devotion, see High, "Projects of devotion."

15. The print version of the article was entitled, "Energy Riches in Bolivia Lure Global Suitors."

16. Krauss, "Green-Energy Race Draws an American Underdog to Bolivia's Lithium."

17. "Green-Energy Race Draws an American Underdog to Bolivia's Lithium."

18. "Green-Energy Race Draws an American Underdog to Bolivia's Lithium."

19. "Green-Energy Race Draws an American Underdog to Bolivia's Lithium."

20. As of December 2024, the publicity video was still available at the following link: https://drive.google.com/file/d/1Sx5NBmdT_bmAoSVTJAR-4ylGaO-aGg7ox/view.

CONCLUSION

1. Kaup, "Negotiating through Nature."
2. Lorde, "The Master's Tools Will Never Dismantle the Master's House."
3. Howe, *Ecologics*, 195; Boyer, *Energopolitics*, 198.
4. Tsing, *The Mushroom at the End of the World*.
5. Anthias, "Contested Energy Futurities at Bolivia's Hydrocarbon-Conservation Frontier."

REFERENCES

Abelvik-Lawson, Helle. "Indigenous Environmental Rights, Participation and Lithium Mining in Argentina and Bolivia: A Socio-Legal Analysis." PhD, University of Essex, 2019.

Abram, Simone. "Problem Solving as Selective Blindness." *Critique of Anthropology* 44, 3 (2024): 256–275.

Ahmad, Samar. "The Lithium Triangle: Where Chile, Argentina, and Bolivia Meet." *Harvard International Review* 41, no. 1 (2020): 51–53.

Almaraz Paz, Sergio. *Réquiem para una república*. La Paz: Universidad Mayor de San Andrés, 1969.

Andreucci, Diego, and Isabella M. Radhuber. "Limits to 'Counter-Neoliberal' Reform: Mining Expansion and the Marginalisation of Post-Extractivist Forces in Evo Morales's Bolivia." *Geoforum* 84 (2017): 280–91.

ANF (Agencia de Noticias Fides). "Pobladores del sudoeste potosino declaran emergencia y rechazan a nuevo gerente de la estatal de litio." January 10, 2020. https://www.noticiasfides.com/economia/pobladores-del-sudoeste-potosino-se-declaran-en-emergencia-y-rechazan-a-nuevo-gerente-de-la-estatal-de-litio--403225#google_vignette.

Antelo Gutiérrez, Sergio. *Los Cambas: Nación sin estado*. Santa Cruz: Instituto de Ciencia, Economía, Educación y Salud (ICEES), 2017.

Anthias, Penelope. *Limits to Decolonization: Indigeneity, Territory, and Hydrocarbon Politics in the Bolivian Chaco*. Ithaca: Cornell University Press, 2018.

———. "The Pluri-Extractivist State: Regional Autonomy and the Limits of Indigenous Representation in Bolivia's Gran Chaco Province." *Journal of Latin American Studies* 54, no. 1 (2022): 125–154.

———. "Contested Energy Futurities at Bolivia's Hydrocarbon-Conservation Frontier." Contesting Transitions: New Directions in the Anthropology of Energy, Environmental Justice, and Resource Imaginaries conference, University of Lausanne, June 19, 2023.

Argento, Melisa. "Espejo de sal: Estructuras de la acción colectiva e integración territorial del proyecto de extracción e industrialización del litio en Bolivia." *Estado & comunes* 2, no. 7 (2018): 227–248.

Arze Vargas, Carlos. "Abandono del discurso estatista en la industrialización del litio." *Perspectiva energética* 17 (2018): 1–3.

Ayub, Mahmood Ali, and Hideo Hashimoto. *The Economics of Tin Mining in Bolivia*. Washington, DC: World Bank, 1985.

Babidge, Sally. *Groundwater Politics. Advanced Extractivism and Slow Resistance*. New York: Berghahn Books, 2025.

Babidge, Sally, and Paola Bolados. "Neoextractivism and Indigenous Water Ritual in Salar de Atacama, Chile." *Latin American Perspectives* 45, no. 5 (2018): 170–185.

Bakker, Karen, and Gavin Bridge. "Material Worlds Redux: Mobilizing Materiality within Critical Resource Geography." In *The Routledge Handbook of Critical Resource Geography*, edited by Matthew Himley, Elizabeth Havice and Gabriela Valdivia, 43–56. London: Routledge, 2021.

Balch, Oliver. "Punta Del Este: Is Uruguay's Uber-Rich 'Gated City' a Glimpse of Our Urban Future?" *Guardian*, January 20, 2016. https://www.theguardian .com/cities/2016/jan/20/punta-del-este-uruguay-elitest-gated-city-urban-future-st-tropez-south-america.

Ballestero, Andrea. *A Future History of Water*. Durham: Duke University Press, 2019.

Barandiarán, Javiera. "Lithium and Development Imaginaries in Chile, Argentina and Bolivia." *World Development* 113 (2019): 381–391.

Bebbington, Anthony, ed. *Social Conflict, Economic, Development and Extractive Industry: Evidence from South America*. New York: Routledge, 2012.

Bednarski, Lukasz. *Lithium: The Global Race for Battery Dominance and the New Energy Revolution*. London: Hurst, 2021.

Benson, Peter. *Tobacco Capitalism: Growers, Migrant Workers, and the Changing Face of a Global Industry*. Princeton: Princeton University Press, 2012.

Black, Megan. *The Global Interior: Mineral Frontiers and American Power*. Cambridge, MA: Harvard University Press, 2018.

Blair, James J. A., Ramón M. Balcázar, Javiera Barandiarán, and Amanda Maxwell. "The 'Alterlives' of Green Extractivism: Lithium Mining and Exhausted Ecologies in the Atacama Desert." *International Development Policy/Revue internationale de politique de développement* 16 (2023). https://doi.org/ https://doi.org/10.4000/poldev.5284. http://journals.openedition.org/poldev/5284.

Blair, James J. A., Noel Vineyard, Dustin Mulvaney, Alida Cantor, Ali Sharbat, Kate Berry, Elizabeth Bartholomew, and Ariana Firebaugh Ornelas. "Lithium and Water: Hydrosocial Impacts across the Life Cycle of Energy Storage." *WIREs Water* (2024). https://doi.org/https://doi.org/10.1002/wat2.1748.

Bonelli, Cristobal, and Martina Gamba. "Underground Roots for Ancestral Futures: Exploring Lithium through an Experimental Alliance between Chemistry and Anthropology." *Science, Technology, and Human Values* (2024). https://doi.org/10.1177/01622439241278377.

Bonelli, Cristobal, and Antonia Walford, eds. *Environmental Alterities*. Manchester: Mattering Press, 2021. https://doi.org/10.28938/97819127291.

Bonelli, Cristobal, Marina Weinberg, and Pablo Ampuero Ruiz. "El litio, un (des) estabilizador de transiciones bipolares." *LASA Forum* 53, no. 1 (2021): 37–43.

Bonneuil, Christophe, and Patrick Petitjean. "Science and French Colonial Policy: Creation of the Orstom; From the Popular Front to the Liberation via Vichy, 1936–1943." In *Science and Technology in a Developing World*, edited by Terry Shinn, Jack Spaapen, and Venni Krishna, 129–78. Berlin: Springer Dordrecht, 1997.

Borsa, Adrian A., Helen A. Fricker, Bruce G. Bills, Jean-Bernard Minster, Claudia C. Carabajal, and Katherine J. Quinn. "Topography of the Salar de Uyuni, Bolivia from Kinematic GPS." *Geophysical Journal International* 172, no. 1 (2008): 31–40.

Bowman, Isaiah. "Results of an Expedition to the Central Andes." *Bulletin of the American Geographical Society* 46, no. 3 (1914): 161–183.

Boyer, Dominic. *Energopolitics: Wind and Power in the Anthropocene*. Durham: Duke University Press, 2019.

Bray, Carol. "KU's First Woman of Chemistry, Mary Elvira Weeks: A History of Our Historian." Department of Chemistry, University of Kansas (1999): 1–5. http://www.chem.ukans.edu/PressReleases/ElviraWeeks/elvira.htm.

Bridge, Gavin. "Material Worlds: Natural Resources, Resource Geography and the Material Economy." *Geography Compass* 3, no. 3 (2009): 1217–1244.

Brown, Walter A. *Lithium: A Doctor, a Drug, and a Breakthrough*. New York City: Liveright, 2019.

Bustos-Gallardo, Beatriz, Gavin Bridge, and Manuel Prieto. "Harvesting Lithium: Water, Brine and the Industrial Dynamics of Production in the Salar de Atacama." *Geoforum* 119 (2021): 177–189.

Calla, Pamela. "Making Sense of May 24th in Sucre: Toward an Antiracist Legislative Agenda." In *Histories of Race and Racism: The Andes and Mesoamerica from Colonial Times to the Present*, edited by Laura Gotkowitz, 311–317. Durham: Duke University Press, 2011.

Calla, Ricardo, Juan Carlos Montenegro Bravo, Yara Montenegro Pinto, and Pablo Poveda Ávila. *Un presente sin futuro: El proyecto de industrialización del litio en Bolivia*. La Paz: CEDLA, 2014.

Carter, William D. "ERTS-A: A New Apogee for Mineral Finding." *Mining Engineering* 23 (1971): 51–53.

Carter, William D., and Lawrence C. Rowan. "Applying Satellite Technology to Energy and Mineral Exploration." *Episodes: Journal of International Geoscience* 1, no. 4 (1978): 19–27.

Choque Canqui, Roberto, and Esteban Ticona Alejo. *Jesús de Machaqa: La marka rebelde, volumen 2 (sublevación y masacre de 1921)*. La Paz: CIPCA/CEDOIN, 1996.

Cole, Jeffrey. *The Potosí Mita, 1573–1700: Compulsory Indian Labor in the Andes*. Stanford: Stanford University Press, 1985.

Cushman, Gregory T. *Guano and the Opening of the Pacific World: A Global Ecological History*. Cambridge: Cambridge University Press, 2013.

Daggett, Cara New. "Renewable Masculinities." University of Lausanne, Lausanne, May 15, 2023.

Davis, Heather, and Zoe Todd. "On the Importance of a Date, or, Decolonizing the Anthropocene." *ACME: An International Journal for Critical Geographies* 16, no. 4 (2017): 761–780.

Davis, J. R., K. A. Howard, S. L. Rettig, R. L. Smith, G. E. Ericksen, François Risacher, Hugo Alarcon, and Ricardo Morales. *Progress Report on Lithium-Related Geologic Investigations in Bolivia.* US Geological Survey (Washington, DC, 1982).

Daza, Weimar Giovanni Iño. "Historia del extractivismo del litio en Bolivia. El movimiento cívico de Potosí y la defensa de los recursos evaporíticos del Salar de Uyuni (1987–1990)." *RevIISE—Revista de Ciencias Sociales y Humanas* 10, no. 10 (2017): 173–88.

De Moore, Greg, and Ann Westmore. *Finding Sanity: John Cade, Lithium, and the Taming of Bipolar Disorder.* Sydney: Allen & Unwin, 2016.

Destrée, Pauline. "Gaseous Politics: Contradictions and Moral Frontiers of the Energy Transition in Ghana." *Critique of Anthropology* 44, no. 3 (2024): 235–255.

Djukanović, Nina. "How Does an Extractivist Frontier Come to Being? The Case of Lithium Mining in Serbia." *East European Politics and Societies* 38, no. 4 (forthcoming).

Donovan, Tristan. *Fizz: How Soda Shook up the World.* Chicago: Chicago Review Press, 2014.

Dube, Ryan. "Look out, Tesla, There's a Really Tiny Competitor in Your Rearview Mirror." *The Wall Street Journal*, April 26, 2023. https://www.wsj.com/articles/electric-vehicle-tiny-bolivia-86b28ebf.

Dunlap, Alexander, and Mariana Riquito. "Social Warfare for Lithium Extraction? Open-Pit Lithium Mining, Counterinsurgency Tactics and Enforcing Green Extractivism in Northern Portugal." *Energy Research & Social Science* 95 (2023). https://doi.org/10.1016/j.erss.2022.102912.

Ericksen, George E., James D. Vine, and Raul Ballón. *Lithium-Rich Brines at Salar de Uyuni and Nearby Salars in Southwestern Bolivia.* Washington, DC: US Geological Survey, 1977.

European Commission. *A European Green Deal: Striving to Be the First Climate-Neutral Continent.* (2020). https://commission.europa.eu/strategy-and-policy/priorities-2019–2024/european-green-deal_en.

Fabricant, Nicole. "Performative Politics: The Camba Countermovement in Eastern Bolivia." *American Ethnologist* 36, no. 4 (2009): 768–783.

Fabricant, Nicole, and Nancy Postero. "Contested Bodies, Contested States: Performance, Emotions, and New Forms of Regional Governance in Santa Cruz." *Journal of Latin American and Caribbean Anthropology* 18, no. 2 (2013): 187–211.

Farthing, Linda, and Thomas Becker. *Coup: A Story of Violence and Resistance in Bolivia.* Chicago: Haymarket Books, 2021.

Fischer, Edward F., and Peter Benson. *Broccoli and Desire: Global Connections and Maya Struggles in Postwar Guatemala*. Stanford: Stanford University Press, 2006.

Fornillo, Bruno. "La energía del litio en Argentina y Bolivia: Comunidad, extractivismo y posdesarrollo." *Colombia Internacional* 93 (2018): 179–201.

Fraser, Nancy. "From Redistribution to Recognition? Dilemmas of Justice in a 'Post-Socialist' Age." *New Left Review* 212, no. June/August (1995): 68–93.

Galeano, Eduardo. *Open Veins of Latin America: Five Centuries of the Pillage of a Continent*. New York City: Monthly Review Press, 1973.

García Linera, Álvaro. "Un Potosí federal dejará de recibir Bs 3.900 millones del Estado." La Paz: Ministerio de Comunicación, 2015.

Geddes, Charles F. *Patiño: Rey del Estaño*. Madrid: A. G. Grupo, 1984.

GNRE (Gerencia Nacional de Recursos Evaporíticos). *Memoria 2015*. La Paz, Bolivia: Unidad de Comunicación GNRE, 2015.

Göbel, Barbara. "La minería del litio en la Puna de Atacama: Interdependencias transregionales y disputas locales." *Iberoamericana (2001-)* 13, no. 49 (2013): 135–149.

Goodale, Mark. "Reclaiming Modernity: Indigenous Cosmopolitanism and the Coming of the Second Revolution in Bolivia." *American Ethnologist* 33, no. 4 (2006): 634–649.

———. *A Revolution in Fragments: Traversing Scales of Justice, Ideology, and Practice in Bolivia*. Durham: Duke University Press 2019.

———. "God, Fatherland, Home: Revealing the Dark Side of Our Anthropological Virtue." *Journal of the Royal Anthropological Institute* 26, no. 2 (2020): 343–64.

———. "Timerendering: Reflections on Chronopolitical Praxis in Bolivia." *Journal of the Royal Anthropological Institute* 28, 3 (2022): 788–806.

———. "An Anthropology of Impossible Futures." *Anthropology Today* 39, no. 5 (2023): 1–2.

———. "Futures Fixed and Foreclosed." *Social Research* 90, no. 4 (2023): 705–24.

———. "Of Crystals and Semiotic Slippage: Lithium Mining, Energy Ambitions, and Extractive Politics in Bolivia." *Anthropological Quarterly* 97, no. 2 (2024): 361–386.

———. "Lithium Scale-Making and Extractivist Counter-Futurities in Bolivia." *Critique of Anthropology* 44, no. 3 (2024): 381–399.

———. "Moving the Bones: Necroextractivism and Capitalist Anti-politics in Bolivia." Unpublished manuscript.

Gordillo, Gastón. *Rubble: The Afterlife of Destruction*. Durham: Duke University Press, 2014.

Gotkowitz, Laura. *A Revolution for Our Rights: Indigenous Struggles for Land and Justice in Bolivia, 1880–1952*. Durham: Duke University Press, 2007.

Grisaffi, Thomas. "A Brief History of Coca: From Pre-Columbian Trade to the Cocaine Economy." In *The Struggle for Natural Resources: Findings from Bolivian History*, edited by Carmen Soliz and Rossana Barragán, 197–235. Albuquerque: University of New Mexico Press, 2024.

Gudynas, Eduardo. "Diez tesis urgentes sobre el nuevo extractivismo. Contextos y demandas bajo el progresismo sudamericano actual." In *Extractivismo, política y sociedad*, edited by Francisco Rhon Dávila, 187–225. Quito: Centro Andino de Acción Popular and Centro Latinoamericano de Ecología Social, 2009.

Gustafson, Bret. "Spectacles of Autonomy and Crisis: Or, What Bulls and Beauty Queens Have to Do with Regionalism in Eastern Bolivia." *Journal of Latin American and Caribbean Anthropology* 11, no. 2 (2006): 351–379.

———. *Bolivia in the Age of Gas.* Durham: Duke University Press, 2020.

Gusterson, Hugh. "Studying up Revisited." *Political and Legal Anthropology Review* 20, no. 1 (1997): 114–119.

Gutiérrez Aguilar, Raquel. *Rhythms of the Pachakuti: Indigenous Uprising and State Power in Bolivia.* Durham: Duke University Press, 2014.

Hand, Eric. "The Salt Flat with Curious Curves." *Nature* (2007). https://www.nature.com/news/2007/071130/full/news.2007.315.html.

Haraway, Donna. "Anthropocene, Capitalocene, Plantationocene, Chthulucene: Making Kin." *Environmental Humanities* 6, no. 1 (2015): 159–165.

Harvey, David. *The Condition of Postmodernity: An Enquiry into the Origins of Cultural Change.* Malden, MA: Blackwell, 1990.

Hayden, Cori. *The Spectacular Generic: Pharmaceuticals and the Simipolitical in Mexico.* Durham: Duke University Press, 2023.

Hecht, Gabrielle. "Interscalar Vehicles for an African Anthropocene: On Waste, Temporality, and Violence." *Cultural Anthropology* 33, no. 1 (2018): 109–141.

Heredia, Florencia, Agostina L. Martinez, and Valentina Surraco Urtubey. "The Importance of Lithium for Achieving a Low-Carbon Future: Overview of the Lithium Extraction in the 'Lithium Triangle'." *Journal of Energy & Natural Resource Law* 38, no. 3 (2020): 213–36.

High, Mette M. "Projects of Devotion: Energy Exploration and Moral Ambition in the Cosmoeconomy of Oil and Gas in the Western United States." *Journal of the Royal Anthropological Institute* 25, no. S1 (2019): 29–46.

———. "Resisting Energy Transitions: Innovation as a Moral Trope in the US Oil and Gas Industry." *Critique of Anthropology* 44, no. 3 (2024): 219–234.

Honneth, Axel. *The Struggle for Recognition: The Moral Grammar of Social Conflicts.* Cambridge, MA: The MIT Press, 1996.

Houston, John, Andrew Butcher, Peter Ehren, Keith Evans, and Linda Godfrey. "The Evaluation of Brine Prospects and the Requirement for Modifications to Filing Standards." *Economic Geology* 106, no. 7 (2011): 1225–1239.

Howe, Cymene. *Ecologics: Wind and Power in the Anthropocene.* Durham: Duke University Press, 2019.

Howe, Cymene and Dominic Boyer. "Aeolian Extractivism and Community Wind in Southern Mexico." *Public Culture* 28, no. 2 (2016): 215–235.

Ingold, Tim. *Being Alive: Essays on Movement, Knowledge and Description.* 2nd ed. Abingdon, Oxon: Routledge, 2022.

IPCC (Intergovernmental Panel on Climate Change). *Special Report: Global Warming of 1.5 °C* (2018). https://www.ipcc.ch/sr15/.

Jemio, Miriam. "Una Bolivia en transición política repiensa cómo industrializar su litio." *Dialogue Earth* (London), May 19, 2020. https://dialogue.earth/es/negocios/35423-una-bolivia-en-transicion-politica-repiensa-como-industrializar-su-litio/.

Jerez Henríquez, Barbara, Ingrid Garcés, and Robinson Torres. "Lithium Extractivism and Water Injustices in the Salar de Atacama, Chile: The Colonial Shadow of Green Electromobility." *Political Geography* 87 (2021). https://doi.org/10.1016/j.polgeo.2021.102382.

Jobson, Ryan Cecil, Macarena Gómez-Barris, Cymene Howe, and Mareike Winchell. "Extractivism's Limits: A Conversation." *Journal of Latin American and Caribbean Anthropology* (2024): 1–6.

Kaup, Brent Z. "Negotiating through Nature: The Resistant Materiality and Materiality of Resistance in Bolivia's Natural Gas Sector." *Geoforum* 39, no. 5 (2008): 1734–1342.

Kingsbury, Donald V. "Lithium's Buzz: Extractivism between Booms in Bolivia, Argentina, and Chile." *Cultural Studies* 37, no. 4 (2022): 580–604.

Kirsch, Stuart. *Mining Capitalism: The Relationship between Corporations and their Critics.* Berkeley: University of California Press, 2014.

———. "Running Out? Rethinking Resource Depletion." *The Extractive Industries and Society* 7, no. 3 (2020): 838–840.

———. "Concrete and the Postcarbon Transition." Unpublished manuscript.

Klein, Herbert S. *Bolivia: The Evolution of a Multi-Ethnic Society.* New York: Oxford University Press, 1982.

Köppel, Jonas. "Lithium Trajectories in Bolivia and Beyond: Encounters against the Supply Chain." PhD, The Graduate Institute of International and Development Studies, 2023.

Krauss, Clifford. "Green-Energy Race Draws an American Underdog to Bolivia's Lithium." *The New York Times*, December 16, 2021. https://www.nytimes.com/2021/12/16/business/energy-environment/bolivia-lithium-electric-cars.html#.

Kunasz, Ihor. *The Lithium Legacy.* Singapore: Jenny Stanford, 2023.

La Lone, Mary B., and Darrell E. La Lone. "The Inka State in the Southern Highlands: State Administrative and Production Enclaves." *Ethnohistory* 34, no. 1 (1987): 47–62.

Lallemand, Alain. "Il est soupçonné de trafic d'acide à destination des narcos: Libération du patron belge 'le plus haut du monde.'" *Le Soir* (Brussels), December 28, 2000. https://www.lesoir.be/art/bolivie-il-est-soupconne-de-trafic-d-acide-a-destinatio_t-20001228-Z0K2QL.html.

Larson, Brooke. *Cochabamba, 1550–1900: Colonialism and Agrarian Transformation in Bolivia.* Durham: Duke University Press, 1988.

Lazar, Sian. *El Alto, Rebel City: Self and Citizenship in Andean Bolivia.* Durham: Duke University Press, 2008.

Li, Tania "Indigeneity, Capitalism, and the Management of Dispossession." *Current Anthropology* 51, no. 3 (2010): 385–414.

Li, Tania, and Pujo Semedi. *Plantation Life: Corporate Occupation in Indonesia's Oil Palm Zone.* Durham: Duke University Press, 2021.

López, Andrés, Martín Obaya, Paulo Pascuini, and Adrián Ramos. *Litio en la Argentina: Oportunidades y desafíos para el desarrollo de la cadena de valor.* Washington, DC: Inter-American Development Bank, 2019.

Lorde, Audre. "The Master's Tools Will Never Dismantle the Master's House." In *This Bridge Called My Back: Writings by Radical Women of Color*, edited by Cherríe Moraga and Gloria Anzaldúa, 94–101. New York City: Kitchen Table Press, 1983.

Los Tiempos. "Pumari convoca a detractores que quieran insultarle: 'Estaré a las 12:00 en la Catedral.'" October 21, 2020. https://www.lostiempos.com/actualidad/pais/20201021/pumari-convoca-detractores-que-quieran-insultarle-estare-1200-catedral.

———. "Pumari pide disculpas entre gritos, insultos y tomatazos." October 22, 2020. https://www.lostiempos.com/actualidad/pais/20201022/potosi-pumari-pide-disculpas-gritos-insultos-tomatazos.

Lowe, Jaime. "I Don't Believe in God, but I Believe in Lithium." *The New York Times*, June 25, 2015. https://www.nytimes.com/2015/06/28/magazine/i-dont-believe-in-god-but-i-believe-in-lithium.html#:~:text=I%20worry%20that%20without%20lithium,but%20I%20believe%20in%20lithium.

Marcus, George. "Ethnography in/of the World System: The Emergence of Multi-Sited Ethnography." *Annual Review of Anthropology* 24, no. 1 (1995): 95–117.

Marston, Andrea. "Alloyed Waterscapes: Mining and Water at the Nexus of Corporate Social Responsibility, Resource Nationalism, and Small-Scale Mining." *Wiley Interdisciplinary Water Review* 4, no. 1 (2016). https://doi.org/10.1002/wat2.1175.

———. "Subsoil Politics: Extraction, Nationalism, and Protest in Bolivia and Peru." *Latin American Perspectives* 45, no. 5 (2018): 229–231.

———. "Strata of the State: Resource Nationalism and Vertical Territory in Bolivia." *Political Geography* 74 (2019). https://doi.org/https://doi.org/10.1016/j.polgeo.2019.102040.

———. *Subterranean Matters: Cooperative Mining and Resource Nationalism in Plurinational Bolivia.* Durham: Duke University Press, 2024.

McNeish, John-Andrew. *Sovereign Forces: Everyday Challenges to Environmental Governance in Latin America.* New York: Berghahn Books, 2021.

McNelly, Angus. *Now We Are in Power: The Politics of Passive Revolution in Twenty-First-Century Bolivia.* Pittsburgh: University of Pittsburgh Press, 2023.

Merry, Sally Engle. *Human Rights and Gender Violence: Translating International Law into Local Justice.* Chicago: University of Chicago Press, 2006.

Mesa, José de, Teresa Gisbert, and Carlos D. Mesa Gisbert. *Historia de Bolivia.* 7th ed. La Paz: Editorial Gisbert y Cía, 2008.

Mintz, Sidney W. *Sweetness and Power: The Place of Sugar in Modern History.* New York: Penguin Books, 1985.

Mitchell, Timothy. *Carbon Democracy: Political Power in the Age of Oil.* London: Verso, 2011.

Mitra, Taj, and Michael Nienaber. "In the New Lithium 'Great Game,' Germany Edges Out China in Bolivia." *Reuters*, January 28, 2019. https://www.reuters.com/article/idUSKCN1PM1LS/.

Molina Rivero, Ramiro. *De memorias e identidades: Los aymaras y urus del sur de Oruro*. La Paz: Instituto de Estudios Bolivianos, Agencia Sueca de Desarrollo Internacional, Fundación Diálogo, 2006.

Montenegro Bravo, Juan Carlos. "El modelo de industrialización del litio en Bolivia." *Revista de ciencias sociales* 34 (2018): 69–82.

Mulvaney, Dustin. *Solar Power: Innovation, Sustainability, and Environmental Justice*. Berkeley: University of California Press, 2019.

Murray, Charles J. *Long Hard Road: The Lithium-Ion Battery and the Electric Car*. West Lafayette, Indiana: Purdue University Press, 2022.

———. "Who Really Invented the Rechargeable Lithium-Ion Battery?" *IEEE Spectrum* (July 23, 2023). https://spectrum.ieee.org/lithium-ion-battery-2662487214.

Nacif, Federico. "Bolivia y el plan de industrialización del litio: un reclamo histórico." *Revista del Centro Cultural de la Cooperación* 5, nos. 14–15 (2012). https://www.centrocultural.coop/revista/1415/bolivia-y-el-plan-de-industrializacion-del-litio-un-reclamo-historico.

Nader, Laura. "Up the Anthropologist: Perspectives Gained from Studying Up." In *Reinventing Anthropology*, edited by Dell Hymes, 284–311. New York: Vintage Books, 1974 [1969].

NASA. "Landsat Science: Data" (2023). https://landsat.gsfc.nasa.gov/data/.

Nash, June. *We Eat the Mines and the Mines Eat Us: Dependency and Exploitation in Bolivian Tin Mines*. New York: Columbia University Press, 1979.

Niezen, Ronald. *The Origins of Indigenism: Human Rights and the Politics of Identity*. Berkeley: University of California Press, 2003.

Nixon, Rob. *Slow Violence and the Environmentalism of the Poor*. Cambridge, MA: Harvard University Press, 2011.

NobelPrize.org. "The Nobel Prize in Chemistry." News release, 2019. https://www.nobelprize.org/prizes/chemistry/2019/prize-announcement/.

Obaya, Martín. *Estudio de caso sobre la gobernanza del litio en el Estado Plurinacional de Bolivia*. Santiago, Chile: CEPAL/United Nations, 2019.

Obaya, Martín, Andrés López, and Paulo Pascuini. "Curb Your Enthusiasm. Challenges to the Development of Lithium-Based Linkages in Argentina." *Resources Policy* 70 (2021). https://doi.org/https://doi.org/10.1016/j.resourpol.2020.101912.

Olivera, Manuel. *La industrialización del litio en Bolivia: Un proyecto estatal y los retos de la gobernanza, el extractivismo histórico y el capital internacional*. La Paz: UNESCO, 2017.

Olivera, Oscar. *¡Cochabamba! Water War in Bolivia*. Cambridge, MA: South End Press, 2004.

Ong, Aihwa. *Flexible Citizenship: The Cultural Logics of Transnationality*. Durham: Duke University Press, 1999.

Orellana Rocha, Walter. "El litio: Una perspectiva fallida para Bolivia." MS, Universidad de Chile, 1995.

Palsson, Gisli, and Heather Anne Swanson. "Down to Earth: Geosocialities and Geopolitics." *Environmental Humanities* 8, no. 2 (2016): 149–71.

Parsons, Keith M., and Robert A. Zaballa. *Bombing the Marshall Islands: A Cold War Tragedy.* Cambridge: Cambridge University Press, 2017.

Plata, Wilfredo. "El discurso autonomista de las élites de Santa Cruz." In *Los barones del Oriente: El poder en Santa Cruz ayer y hoy*, edited by Ximena Soruco, Wilfredo Plata, and Gustavo Medeiros, 101–71. Santa Cruz: Fundación Tierra, 2008.

Poveda, Pablo. *Industrialización del litio en Bolivia: ¿100% estatal? Perspectiva energética* 22 (2018): 1–2.

Prada, Paulo. "Bolivian Nationalizes the Oil and Gas Sector." *The New York Times*, May 2, 2006. https://www.nytimes.com/2006/05/02/world/americas/02bolivia.html.

Quezada Cortez, Guido and Nelson Carvajal Velasco. *Salar de Uyuni: Geología y recursos de litio del depósito evaporítico más grande del mundo.* La Paz: Imprenta Megaprint, 2020.

Quispe, Aline. "Hasta 2019, Bolivia invertirá $us 925 mm en la industria del litio." *La Rázon* (La Paz), August 17, 2015.

Revette, Anna C. "This Time it's Different: Lithium Extraction, Cultural Politics and Development in Bolivia." *Third World Quarterly* 38, no. 1 (2016): 149–68.

Riofrancos, Thea. *Resource Radicals: From Petro-Nationalism to Post-Extractivism in Ecuador.* Durham: Duke University Press, 2020.

———. *Extraction: The Frontiers of Green Capitalism.* New York: W. W. Norton, 2025.

Risacher, François. *Estudio económico del Salar de Uyuni.* La Paz: ORSTOM, 1989.

Risacher, François and Bertrand Fritz. "Bromine Geochemistry of Salar de Uyuni and Deeper Salt Crusts, Central Altiplano, Bolivia." *Chemical Geology* 167, no. 3–4 (2000): 373–392.

———. "Origin of Salts and Brine Evolution of Bolivian and Chilean Salars." *Aquatic Geochemistry* 15, no. 1–2 (2009): 123–157.

Robbins, Joel. "The Present and the Future in the Present: Religion, Values and Climate Change." Departmental Seminar in Anthropology, University of Oxford, December 1, 2023.

Rodríguez-Carmona, Antonio, and Iván Aranda Garoz. *De la salmuera a la batería, soberanía y cadenas de valor, un balance de la política de industrialización minera del Gobierno del MAS (2006–2013).* La Paz: Centro de Investigaciones Sociales/PNUD, 2014.

Romero, Hugo, Manuel Méndez, and Pamela Smith. "Mining Development and Environmental Injustice in the Atacama Desert of Northern Chile." *Environmental Justice* 5, no. 2 (2012): 70–76.

Sanchez-Lopez, Maria Daniela. "From a White Desert to the Largest World Deposit of Lithium: Symbolic Meanings and Materialities of the Uyuni Salt Flat in Bolivia." *Antipode* 51, no. 4 (2019): 1318–1339.

———. "Territory and Lithium Extraction: The Great Land of Lipez and the Uyuni Salt Flat in Bolivia." *Political Geography* 90 (2021). https://doi.org/10.1016/j. polgeo.2021.102456.

Sassen, Saskia. *Expulsions: Brutality and Complexity in the Global Economy.* Cambridge: Harvard University Press, 2014.

Schiaffini, Hernán. "Litio, llamas y sal en la Puna Argentina: pueblos originarios y expropiación en torno al control territorial de Salinas Grandes." *Entremados y Perspectivas* 3, no. 3 (2013): 121–136.

Schröter, David Luis. "The Future Is Now: The Role of Competing Future Imaginaries in Activist Responses to Lithium Mining in Bolivia." PhD, University of Lausanne, 2025.

Seefeldt, Jennapher Lunde. "Lessons from the Lithium Triangle: Considering Policy Explanations for the Variation in Lithium Industry Development in the 'Lithium Triangle' Countries of Chile, Argentina, and Bolivia." *Politics & Policy* 48, no. 4 (2020): 727–765.

Servant, Michel, and Jean-Charles Fontes. "Les lacs quaternaires des hauts plateaux des Andes boliviennes: Premières interprétations paléoclimatiques." *Cahiers OSTOM, Série Géologie* 10, no. 1 (1978): 9–23.

Shorter, Edward. "The History of Lithium Therapy." *Bipolar Disorders: An International Journal of Psychiatry and Neurosciences* 11, no. S2 (2009): 4–9.

Smith, Jessica M. *Extracting Accountability: Engineers and Corporate Social Responsibility.* Cambridge: The MIT Press, 2021.

Smith, Neil. *American Empire: Roosevelt's Geographer and the Prelude to Globalization.* Berkeley: University of California Press, 2004.

Solón Romero Peredo, José Carlos. *Espejismos de abundancia: Los mitos de la industrialización del litio en el Salar de Uyuni.* La Paz: Editorial Plural, 2022.

Soto Hernandez, Daniela, and Peter Newell. "Oro blanco: Assembling Extractivism in the Lithium Triangle." *The Journal of Peasant Studies* 49, no. 5 (2022): 945–968.

Stoertz, George E., and William D. Carter. "Hydrogeology of Closed Basins and Deserts of South America." In *ERTS-1: A New Window on Our Planet*, edited by Richard S. Williams and William D. Carter, 76–80. Washington, DC: US Department of the Interior, 1976.

Svampa, Maristella. *Neo-Extractivism in Latin America: Socio-Environmental Conflicts, the Territorial Turn, and New Political Narratives.* Cambridge: Cambridge University Press, 2019.

Szeman, Imre. *After Oil.* Edmonton: Petrocultures Research Group, 2016.

———. *On Petrocultures: Globalization, Culture, and Energy.* Morgantown: West Virginia University Press, 2019.

Szeman, Imre, and Dominic Boyer, eds. *Energy Humanities: An Anthology.* Baltimore: Johns Hopkins University Press, 2017.

Szeman, Imre, and Jennifer Wenzel, eds. *Energized: Keywords for a New Politics of Energy.* Morgantown: West Virginia University Press, 2024.

Szoke, Zsofia J. "The Lithium Economy: Bolivia's 'New' Resource and Its Role in Revolutionary Politics." PhD, University of New Mexico, 2023.

Tellería, Loreta. "La hegemonía verde: Control de los recursos naturales en Bolivia bajo el discurso medioambienta." *La Migraña: Revista de Análisis Político* 14 (2015): 25–41.

Tobar, Héctor. "Revolt on High: The Indians of Bolivia's El Alto Lead a Drive for Social Change That Has Toppled Two Presidents." *Los Angeles Times*, June 16, 2005.

Tsing, Anna Lowenhaupt. *The Mushroom at the End of the World: On the Possibility of Life in Capitalist Ruins.* Princeton: Princeton University Press, 2015.

Twidale, Susana. "'Age of Electricity' to Follow Looming Fossil Fuel Peak, IEA Says." *Reuters*, October 16, 2024. https://www.reuters.com/business/energy/age-electricity-follow-looming-fossil-fuel-peak-iea-says-2024-10-16/.

US Department of the Interior. "Earth's Resources to Be Studied from Space." News release, September 21, 1966.

USGS (US Geological Survey). *Mineral Commodity Summaries (Lithium, 2020).* (2020). https://doi.org/10.3133/mcs2020.

———. *Mineral Commodity Summaries (Lithium, 2022).* (2022). https://doi.org/10.3133/mcs2022.

———. *Mineral Commodity Summaries (Lithium, 2024).* (2024). https://doi.org/10.3133/mcs2024.

Velasco Hurtado, Carlos. "VII congreso nacional de metalurgia y ciencia de materiales." *Revista Metalúrgica UTO* 28 (2007): 31–33.

Vera, María L., Walter R. Torres, Claudia I. Galli, Alexandre Chagnes, and Victoria Flexer. "Environmental Impact of Direct Lithium Extraction from Brines." *Nature Reviews Earth & Environment* 4 (2023): 149–165.

Vice Ministry of Communication. "Presidente Morales inaugura la planta industrial de cloruro de potasio en Uyuni." News release, October 7, 2018. https://www.comunicacion.gob.bo/?q=20181007/26070.

Von Vacano, Diego. "Power over Energy: Lithium, Bolivia, and the Global Struggle for the New Green Economy." Unpublished manuscript.

Voskoboynik, Daniel Macmillen, and Diego Andreucci. "Greening Extractivism: Environmental Discourses and Resource Governance in the 'Lithium Triangle.'" *EPE: Nature and Space* 5, no. 2 (2022): 787–809.

Weeks, Mary Elvira, and Mary E. Larson. "J. A. Arfwedson and His Services to Chemistry." *Journal of Chemical Education* 14, no. 9 (1937): 403–407.

Weszkalnys, Gisa. "Geology, Potentiality, Speculation: On the Indeterminacy of First Oil." *Cultural Anthropology* 30, no. 4 (2015): 611–639.

Whyte, Jessica. *The Morals of the Market: Human Rights and the Rise of Neoliberalism.* London: Verso, 2019.

Whyte, Kyle Powys. "Indigenous Climate Change Studies: Indigenizing Futures, Decolonizing the Anthropocene." *English Language Notes* 55, no. 1–2 (2017): 153–162.

Winchell, Mareike. *After Servitude: Elusive Property and the Ethics of Kinship in Bolivia.* Berkeley: University of California Press, 2022.

———. "Fire's Alter-Lives: Climate Change Adaption and Settler Futurity in Bolivia." *Anthropology News*, August 21, 2024.

World Economic Forum. "IEA: Electric Cars, Clean Energy Policies to Drive Peak Fossil Fuel Demand by 2030." October 24, 2023. https://www.weforum.org /stories/2023/10/iea-energy-peak-fossil-fuel-demand-by-2030/.

Wright, Lawrence. "Lithium Dreams: Can Bolivia Become the Saudi Arabia of the Electric-Car Era?" *New Yorker*, March 22, 2010. https://www.newyorker.com /magazine/2010/03/22/lithium-dreams.

———. "The Astonishing Transformation of Austin." *New Yorker,* February 6, 2023. https://www.newyorker.com/magazine/2023/02/13/the-astonishing-transformation-of-austin.

Xiang, Biao. "Multi-Scalar Ethnography: An Approach for Critical Engagement with

Migration and Social Change." *Ethnography* 14, no. 3 (2013): 282–299.

Yarrow, Thomas. "Remains of the Future: Rethinking the Space and Time of Ruination through the Volta Resettlement Project, Ghana." *Cultural Anthropology* 32, no. 4 (2017): 566–91.

Zuboff, Shoshana. *The Age of Surveillance Capitalism: The Fight for a Human Future at the New Frontier of Power.* New York City: PublicAffairs, 2019.

INDEX

market cap, of Tesla, 11
market capitalism, 46
market dynamics, 146
Márquez, José Carlos, 166–70, 173–74, 176, 180, 182, 190–91
Marshall Islands, 37
Marston, Andrea, 107, 2534n5
MAS. *See* Movement to Socialism
masculinities, renewable, 188
McNeish, John-Andrew, 71
megafires, 18
membranes, 208–9
memory, collective, 62
mental illness, lithium treatments for, 36
mercantile colonialism, 24, 42
Merry, Sally Engle, 19–20
Mesa, Carlos, 62, 81
Mesa, José de, 43
Mesa Gisbert, Carlos, 43, 246n50
Metalin, 166–70, 173, 175
Mette High, 198
Mexico, 132, 136, 255n11
Miami Dolphins, 203, 204
Microelectronics and Computer Consortium, 200
Minchin, Jack "John" B., 98–99
mineral finding, 51
Ministry of Hydrocarbons and Energy, 189
mini-wagons, 167
Minnesota, University of, 89
Mintz, Sidney, 31
Mitsubishi, 67
mobility, electrification of, 11, 21
molecular theory, 210
Molina, Franklin, 158, 214
Molina Ortiz, Franklin, 252n28
Montenegro, Juan Carlos, 88, 90, 113, 140, 147
Morales, Evo, 25, 55, 60, 62–69, 108–12, 125, 219, 222, 252n26; coca growers' union of, 46; Plurinational Constitutional Court and, 74; School of the Americas and, 82; transparency and, 193
Morales, Ricardo, 54
Movement to Socialism (MAS), 25, 62–68, 111, 171, 182, 197–98, 212, 252n26; Arce, L., and, 97; collapse of, 83; constitutional conflict of, 74; "first lithium"

and, 92; gas pipeline deal and, 47; lithium industrialization and, 8, 64, 73; Potosí Department and, 142; productive sovereignty and, 120; Téllez and, 136; War of the Pacific and, 45; Wright on, 8; YLB-ACISA and, 82
Movimiento Indígena Pachakuti, 60, 62
MRI. *See* magnetic resonance imagining
Mullenax, Andrew, 214–16
multiscalar research, 21
Murray, Charles J., 38–39
"Musée des Beaux Art" (Auden), 211
Musk, Elon, 200, 206
"The Myth of Tunupa" (oral tradition), 94

Nagasaki, 36
NASA. *See* National Aeronautics and Space Administration
NASA Goddard Space Flight Center, 51
Nash, June, 107
National Aeronautics and Space Administration (NASA), 50–51, 53, 100
National Faculty of Engineering. See *Facultad Nacional de Ingeniería*
National Football League (NFL), 203, 205
nationalism: Camba ethnonationalism, 68, 248n15; resource, 57; scientific, 130, 151
nationalization: hydrocarbon, 66; of oil and gas industries, 112
national policy, 104
National Revolution, 170
natural gas, 25, 62–63; nationalization, 112; pipeline deal, 47
The Nature and Treatment of Gout and Rheumatic Gout (Garrod), 34
Nazca plate, 95
neocolonialism, 6
neo-extractivism, 15, 17, 97
neoliberal ancien régime (1982–2005), 45
neoliberalization, "shock doctrine," 144
Netherlands, 171
New Economic Policy, 144
New School for Social Research, 89
The New Yorker, 8, 199
New York Times, 65, 217, 218, 222
New Zealand, 99
NFL. *See* National Football League
NFL Players Association, 205

Pumari, Marco, 80–81, 88, 110, 134–35
purification, 225

Qin EV300, 166
Qinghai Institute of Salt Flats, 195
Quantum Batteries, 191, 227
Quantum Motors, 155, 174–90, 176*fig.*, 177*fig.*, 178*fig.*, 179*fig*
Quezada Cortez, Guido, 11–12, 89–90, 102, 109
Quintanilla & Quintanilla, 146
Quiroga Ramírez, Jorge, 46
Quiroga Santa Cruz, Marcelo, 114
Quispe, Felipe ("El Mallku"), 60, 62

radiation exposure, 37
railway networks, 99
rain, 3–4
Ramos Mamani, Carlos Humberto, 120–21, 137, 157–58, 217, 221
Raw Deal (restaurant), 199
La Razón, 25
RCA Records, 205
Reagan Administration, 199
real mining, 167
reciprocal exchange, 3
recovery rate, 194, 209
Red Grasshopper, 153, 162
Regional Minerals Attaché, 52
"remains of the future," 29
renewable masculinities, 188
Repsol, 65
resistance, 16, 96, 224–25
resistant materiality, 224
resource colonialism, 6
resource genealogy, 30
resource geopolitics, 15, 40, 45, 48
resource nationalism, 57
resource regionalism, 53–57
resource sovereignty, 71
Rettig, Shirley L., 51
reverse osmosis, 209
Reyss-Brion, Roger, 34
"Rich Mountain." *See* Cerro Rico
rights: human, 27; Indigenous, 46; of women, 20
Ring of Fire, 95, 101
Rio de Janeiro, Federal University of, 115

Riofrancos, Thea, 243n18
Río Grande delta, 101, 109, 119
Río Grande drainage basin, 54
Rio Tinto, 209
Riquito, Mariana, 16
Risacher, François, 12, 54, 109
risk-taking, 198
rock salt. *See* halite
Roelants du Vivier, Guillaume (Guillermo), 66
Rojas, Nicolas, 144
Roosevelt, Franklin D., 250n7
Rosatom, 195, 214
Royal College of Mines (Sweden), 32
Royal Dutch Shell, 245n48
rubber trees, 245n48
rubble, 29, 58
rural electrification project, 155
Russia, 129, 195, 214, 222

Saavedra, Bautista, 2534n5
Saavedra Villarroel, Roberto, 88–89
Saint Mary's University, 135
Salar de Atacama, 45, 102, 226
Salar de Coipasa, 12, 50, 52
Salar de Empexa, 52
Salar del Hombre Muerto, 226
Salar de Uyuni, xiii*map,* 11–12, 18, 36, 50–52, 97–100, 247n69; basic infrastructure in, 8; brine of, 5, 72, 214; COMIBOL and, 67; construction at, 96*fig.*; FMC-Lithco and, 57; FRUTCAS and, 66; green solvent in, 141; as immature salar, 102; industrial building view, 61*fig.*; lithium industrialization in, 163; Potosí Department and, 134; productive chain and, 159; UMSA and, 53; volcanism in, 101; YLB-ACISA and, 78
salt. *See* sodium
saltpeter. *See* sodium nitrate
salt substitute panic, 34–35
Sánchez de Lozada, Gonzalo (Goni), 46–47, 56, 170, 246n50
San Cristóbal Mine, 1, 3–4, 18
Saudi Arabia, 38, 65
Saudi Aramco, 11
Savannah Resources, 16
scale, 21

Whittingham, M. Stanley, 38–39, 140, 147
Whyte, Jessica, 46
Williams, William Carlos, 6
Wilson, Woodrow, 250n7
winchuka (assassin bug), 175
wind power, 16, 122
WMI. *See* World Manufacturers Identifier
women's rights, globalization of, 20
Wood, Martin, 152
World Bank, 63
World Manufacturers Identifier (WMI),
    254n7
World War I, 245n46, 250n7
World War II, 1, 36, 112, 250n7
Wright, Lawrence, 8, 199

Xinjiang TBEA Group, 249n33

Yale University, 100
Yarrow, Thomas, 29
yatiris (Aymara ritual specialists), 65
"Yes, I am Electric" (car signage), 161f
YLB-ACISA, 78, 82, 87, 92, 110, 134
Yoshino, Akira, 38–39, 140, 147
YPFB, 120

Zaballa, Robert A., 37
Zimpertec, 155
Zona Sur, 192
Zuboff, Shoshana, 52
Zuleta, Juan Carlos, 89

Founded in 1893,
UNIVERSITY OF CALIFORNIA PRESS
publishes bold, progressive books and journals
on topics in the arts, humanities, social sciences,
and natural sciences—with a focus on social
justice issues—that inspire thought and action
among readers worldwide.

The UC PRESS FOUNDATION
raises funds to uphold the press's vital role
as an independent, nonprofit publisher, and
receives philanthropic support from a wide
range of individuals and institutions—and from
committed readers like you. To learn more, visit
ucpress.edu/supportus.